T0319491

THE TENSIONS OF ALGORITHMIC THINKING

Automation, Intelligence and the Politics of Knowing

David Beer

BRISTOL
UNIVERSITY
PRESS

First published in Great Britain in 2023 by

Bristol University Press
University of Bristol
1–9 Old Park Hill
Bristol
BS2 8BB
UK
t: +44 (0)117 374 6645
e: bup-info@bristol.ac.uk

Details of international sales and distribution partners are available at bristoluniversitypress.co.uk

© Bristol University Press 2023

British Library Cataloguing in Publication Data
A catalogue record for this book is available from the British Library

ISBN 978-1-5292-1289-1 hardcover
ISBN 978-1-5292-1291-4 ePub
ISBN 978-1-5292-1292-1 ePdf

Cover design: Namkwan Cho
Front cover image: shutterstock.com
Bristol University Press use environmentally responsible print partners.
Printed and bound in Great Britain by CPI Group (UK) Ltd, Croydon, CR0 4YY

This one is for Nona

Contents

Acknowledgements

First, I'd like to thank the inflammatory bowel disease team based at York Teaching Hospital. I owe them a huge debt of gratitude for their care and expertise. While I was working on this book my wonderful colleague and friend Andrew Webster sadly died. I'd like to remember him and his support here. In addition, I'd like to thank anyone who has talked to me about the ideas I've been working on over the last three years. Thanks go to Paul Stevens at Bristol University Press for supporting the project and for helping me to realize it, especially when I wasn't quite sure of its exact direction in the early stages. And thanks to the anonymous reviewers of the proposal and manuscript; a number of their suggestions were very useful in shaping the content and direction of the book. Thanks go to Cupcake, Orion and Dexter for the companionship. As ever, extra special thanks go to Erik and Martha.

Introduction: Tense Thinking and the Myths of an Algorithmic New Life

Without wanting to sound too epochal, it could be said that we are living in algorithmic times. We may not want to go so far, and I find myself trying to resist the temptation, but it has become hard to draw any other conclusion. The type of 'programmed sociality' to which Taina Bucher (2018: 4) has referred has become impossible to deny, especially as the algorithm 'induces, augments, supports, and produces sociality'. Different types of algorithms have come to have very large-scale social consequences, ranging from shaping what people discover, experience and consume as they go about their mediated lives, through to financial decisions, trading choices, differential treatment around insurance, credit and lending and then on to housing, border controls and a wide array of other processes and decisions that just simply cannot be listed. Such features of the contemporary landscape have led to the compelling conclusion that this is a type of 'algorithmic culture' (Striphas, 2015) cultivated within the structures and relations of an 'algorithmic society' (Peeters and Schuilenburg, 2021). Given these circumstances we may even now be living an 'algorithmic life' (Amoore and Piotukh, 2016). Such conclusions are certainly merited. Part of the reason we might think of these as algorithmic times, if you would let me stick with that slightly hyperbolic phrasing for a moment longer, is just how long the list would be if we were to try to itemize every single way that algorithms have social or individual consequences. And even then, because of their often-invisible roles and the sheer complexity of overlapping systems, the list would be impossible to complete. The algorithm has become too enmeshed in the social world for it to be untangled from it.

Yet, far from being sleek and uncontested technologies informed by mutually recognized ideals or shared notions of progress, these algorithmic times are fraught with tensions. Algorithmic thinking is tense. This book is concerned with elaborating and conceptualizing some of these competing forces. More than this though, this book argues that these algorithmic times can only really be grasped if we are to engage with the specific nature of these strains. Understanding algorithmic times requires a sustained focus upon the tensions of algorithmic thinking. I should add at this opening juncture that when I refer to the term *algorithmic thinking* in the following pages of this book, I mean both the thinking that is done about algorithmic systems *and* the thinking that is done by and through those systems.

Writing in the early 1960s, as part of a broader project on the properties of modernity, Henri Lefebvre (1995) reflected on the increasing importance of the proliferating impressions of *ways of life that were yet to fully arrive*. The changes that were imagined to be just around the corner were, he noted, of particular significance to modernity itself. Such impressions of potential near futures were becoming, so it goes, an integral part of the very experience of modernity. Lefebvre had noticed that there was a feeling of being on the cusp of change and that this apparent threshold brought with it a palpable sense of opportunity and possibility. A strong sense of the change that might be on the way created a space to reshape, to redraw and, possibly, to bring new worlds into existence. The feeling of there being transformations afoot gave, it would seem, a greater sense of investment and intent to the shaping of the present. Lefebvre wondered if these coming ways of life carried a particular significance for understanding the times in which he was writing – it was a time defined by a mix of both the now and the next. The transformations of modernity had created a feeling that more was on the horizon.

Tellingly, Lefebvre (1995: 65) wrote that there 'is something we call a "modern myth" which sees part of reality as an image of what is possible and what is not possible, sometimes illuminating it, sometimes hiding it from our eyes: it is the myth of the new life'. The shifting visibilities of what might be possible, a future set in the now, that is Lefebvre's concern. The sense of possibility in the present is, he argues, part of the way that certain horizons become visible. This *myth of the new life*, Lefebvre claimed, is embedded into modernity and, when it comes into view, becomes something of a focal point, a preoccupation and, therefore, an active part of social life. As such, the myth of the new life becomes a defining presence in shaping what is understood to be possible and what is not. The mythological new life described by Lefebvre is never quite reached; it remains a spectre hovering on the near horizon. It becomes a type of guide to the limits of possibility rather than a fixed material presence.

It may take on the immaterial form of an apparition, yet the myth of the new life is still a potent and very real force in the present. The promise is, Lefebvre (1995: 65) continues, that:

Any day now we would be entering the new life. This life was already there, close at hand, possible, almost a presence, but smothered, sidelined, absent, yet ever-ready for the moment when it would be released into the open. Therefore this new life which was about to spring forth from the world of appearances was itself new in appearance only.

Almost present, the new life remains tantalizingly out of touch. Any day now, as he puts it. The myths have a temporality and proximity to them. The new life is imminent as are its promises of something different, a change that is always just around the corner. These impressions of what might come in the future remain continually in the near distance, just out of reach. For Lefebvre, the promise of *the new life* is a powerful and determining presence. These myths also define the version of modernity that is being lived and the momentum and direction it might have. Implicitly the new life that is imagined, Lefebvre acknowledges, is a 'critique of what already exists'. By setting out the limits it is focused on how they might change or be moved. And so existing shortcomings are implied by the push for change. The promises of transformation are based, essentially and often tacitly, on apparent deficiencies in the present. By suggesting that an alternative is needed, the promises incorporated into the myth of a new life are a gently disruptive presence that keeps the churn of social change moving, perhaps even accelerating it while also setting out its potential directions. We are living in very different times to those that Lefebvre sought to explain, yet we may still wonder how automated systems and algorithmic thinking have such myths of a new life bundled within them. We might then also wonder how those myths come to shape the sense of possibility and the direction of automation. What we can think of as *the myths of an algorithmic new life* are a fundamental part of how algorithmic systems are expanding, taking hold and developing.

When it comes to the presence of algorithms, automation and new forms of intelligence, the ideas around what is possible are prevalent, powerful and have a strong sense of possibility and proximity coded into them. The algorithmic new life is just around the corner, along with all the possibilities it is bringing and will continue to bring. The myths of a new life that accompany automation are accompanied by tangible yet profoundly different visions of the *nearly-here* or the *not-quite-yet* in which automated systems transform the social world in ways that are just starting to be imagined. A transformative force is pictured to be hovering on the near horizon – a type of algorithmic apparition – along with its attendant ideals of what can be achieved and the visions of the lifestyles they will bring. An updated and souped-up version of Lefebvre's myths of a 'new life' continues to boil. One difference we might observe in such current myths of an algorithmic new life is that while they are perhaps more active in embracing the potential unknowns that the future might bring, they also seek to cross, breach and disrupt boundaries in even more active ways. There is an even greater push for those just out of reach forms of social transformation. The defining of possibility has become even more important in the actively disruptive modes of reasoning that accompany algorithmic thinking.

In its current incarnation, algorithmic thinking often has a privileged place within these ongoing visions of a possible 'new life'. Automation is at the fore. In this case, this new life is not just about the alternative ways of living that might be brought about by the expansions and transformations of modernity, it also has a second meaning: the new life of the machines themselves as embodied in their capacity for thinking, decision-making, learning and even autonomy. When thinking about automation we are, like Lefebvre when he was contemplating the march of modernity, seeing a reality in which a notion of a new life is an integral part of the realities of the moment. The movements of automation and algorithmic thinking have this type of new life associated with them, central to them in fact, and it is a myth of a new life in which the promise and potentials are attractors for change, disruption and alteration. In short, they are sites of tension. These myths too are often positioned just, narrowly, out of reach. They will, we get the impression, come any day now. Algorithmic thinking has its spectres and their form is discernible.

Thinking about algorithmic thinking

This brings us to the question of what this automated new life might bring. And, alongside this, we might also wonder at this point how we might see algorithmic thinking as both myth and materiality (a question that has been raised in environmental terms by Crawford, 2021). There are many more questions, not least concerning the tussles over the sense of possibility associated with these developments and how these form into certain types of horizon. Yet among these many questions, in the case of this particular book at least, I want to focus on a particular aspect. The aim here is to think about the tensions that arise from the competing forces at play in the advancement and application of automation of different types. The myths of an algorithmic new life may cohere and they may even become subsumed into the materiality of systems, still they are defined by tensions. This book is concerned with understanding those tensions.

Of course, the ideas behind automated machines and algorithmic thinking already have a long history, as do ideas about how such machines relate to the figure or category of the human (as discussed in detail in chapter 3 of Rees and Sleigh, 2020). And so algorithmic thinking needs to be understood and explored in historical and situated terms. The involvement of automation in the myths of the new life might need us to question the notion of the *new* contained within them. Still, though, the escalation of the availability of data in the last two decades (see Kitchin, 2014) and the growth of value in its analysis and applications (see Beer, 2019a), along with a range of attendant technological developments of various sorts, have sparked new life into the pursuit of automated systems, artificial intelligence (AI), machine learning and a wide range of forms of algorithmic thinking. The many issues that arise here cannot be captured in their entirety, yet a focus on the tensions is one way of elaborating and exploring the key features of algorithmic thinking. Ferrari and Graham (2021: 815) have argued that there are 'fissures in algorithmic power' and that these fissures can be understood as 'moments in which algorithms do not govern as intended'. In line with this observation, we might conclude that there are also fissures within algorithmic thinking. I'm not assuming here that algorithmic thinking is without limits or that it somehow has a pure, containable and singular form, nor am I arguing that algorithmic thinking itself cannot be resisted or developed in ways that go beyond or outside of its intentions. I'm actually hoping to keep these things within the analysis and use their presence to think about the tensions that are created. By foregrounding the tensions, I hope to reflect on these very aspects of algorithmic thinking and animate such fissures in the very idea of the algorithm and in the thinking these systems facilitate.

In the same way that Lefebvre's myths of a new life set out what was possible and what was not, in a context of automated systems what we know and how we know it represent important questions (not least because we might understand knowing itself to have moved out into these very systems, as I will discuss in Chapter 4 in particular). In the conclusion to this book I begin to think about how this drive and desire to expand and develop automation and algorithmic thinking can be thought of in terms of a *will to automate* (see Chapter 6). What is at stake here are the very limits of *the known* on one hand and *the human* on the other. What is difficult is to find ways to look across these limits so as to see their impacts and how, importantly, we might understand, conceptualize and explain them. This book is not thinking in terms of ongoing and immovable tensions that are found at these limits. By generating and developing four key concepts at these limits, which I will describe in a moment, this book tackles the tensions that are at the heart of algorithmic thinking.

Crucially, Luciana Parisi (2019: 93), reflecting on the problem of indeterminacy in computation, has argued that 'a critical view of computation requires an effort to unpack this tension to account for indeterminacy in conditions of knowledge that both constrain and enable the scientific and manifest image of algorithmic thinking'. Parisi was referring in this passage to something very specific, but if we put this aside for the moment then this proposition leaves us with a requirement to try to unpack other tensions around computation, automation and modes of, as she refers to it here, *algorithmic thinking*. That is the aim of this book; it seeks to develop a conceptual repertoire for exploring the competing forces and tensions that are enveloped within *algorithmic thinking*. Parisi also points out that as 'statistics and probability theory enter the realm of AI with learning algorithms in neural networks, new understandings of cognition, logical thinking and reasoning have come to the fore' (Parisi, 2019: 92). This hints at how tensions and indeterminacies are an integral part of the new types of cognition and logic that are espoused. And so, following this, it is with a focus on these tensions that we might propagate further understandings of algorithmic thinking not as a finished product or a fixed moment but as a set of processes defined by competing forces and the tensions they produce. As these systems continue to establish themselves, so social dynamics alter and new tensions emerge – in the same way that technologies are not fixed but adapt and develop so too the tensions associated with them are dynamic. The myths of a new life, as Lefebvre suggested, are not fixed and neither are the possibilities that are carved into them. The same is likely to be the case of the myths of the algorithmic new life. To reiterate, this book explores the tensions that arise as the algorithmically defined new life is mythologized, materialized and established.

The forces and tensions of algorithmic thinking

In their account of the kind of learning that machine learning represents, Tyler Reigeluth and Michael Castelle (2021: 80) make a broader point about the direction in which such studies may wish to head:

> we need a *social theory of machine learning*: i.e. a theory that accounts for the interactive underpinnings and dynamics of artificial 'learning' processes, even those thought to be 'unsupervised.' But in doing so, however, we must be careful not to resort to the term 'social' as a self-explanatory term that is simply opposed to the 'technical'.

Putting the particular focus on machine learning aside for a moment, this passage is a part of a broader call to engage conceptually with the changes that various forms of automation are bringing and especially those changes brought about by advancing forms of machine learning and the like (I'll return to unsupervised learning more directly in Chapters 4 and 5). A theoretical engagement with the issues is what is being suggested in the previous passage and this is what I'm hoping to do, in part at least, with this book. I'm picking up here on the call for theory that deals with the underpinnings and dynamics. The changes around automation, as embodied in developments included under the label machine learning among others, call for new or adapted conceptual resources. The vast changes also call for approaches that seek to identify connections and shared properties across different contexts.

Reigeluth and Castelle's warning about using the term social as a catchall along with their warning about treating the social and technical as distinct remain pertinent and provocative. In many ways it is the technical that is often presented as the main source of uncertainty in these systems, whereas their point is that there may be a tendency towards an underplaying of the uncertainty of social dynamics. Their concern is with a blunt and underdeveloped notion of the social when seeking to understand automation. Perhaps the shock of the new technologies leads to the technical being prioritized in the analysis. Reigeluth and Castelle are not saying this is always the case, rather their concern is that the discourse around machine learning in particular foregrounds the technological hitches and possibilities without necessarily integrating a conceptual encounter with the social context. There is plenty of work that strikes a balance between the social and the technical in the analysis of various types of automation. I will refer to a good deal of this type of work across this book's chapters; still, Reigeluth and Castelle's warning is worth reflecting upon. The social is, of course, never self-explanatory and shouldn't be overlooked as a result of the dazzling and distracting technological formations that are continuing to emerge. Perhaps this is another way of warning against the simple separation between the technical and the social, but I think it goes beyond those types of separations to force a reflection on what the social is when active, performative and potentially thinking technologies become a part of it.

The next step is to think about how to get at the uncertainties of automation. When we speak of automation we are really talking about the ongoing process of automating rather than an end point. Within these automating processes of algorithmic thinking this book identifies and explores two sets of defining and competing forces that generate tensions. To reveal the tensions, the chapters of this book explore and conceptualize the four vectors of these two forces. The book probes at these tensions through specific instances and seeks to build up concepts that operate at each of the four vectors that have been identified. The following four chapters each take one of these forces in turn, exposing aspects of their features and offering resources for thinking about them further and as they continue to mutate. The chapters look at the push and pull around automation, examining how different features and approaches pull in different directions, creating tensions in the process. The vectors and forces that produce and afford the tensions of algorithmic thinking are summarized in Figure 1.1.

On the surface, this may appear to represent a simple set of relations. However, the minimalism in the diagram belies a vast complexity and depth to the tensions these simple lines are being used to represent. Let me break

Figure 1.1: The competing forces that produce and afford the tensions of algorithmic thinking

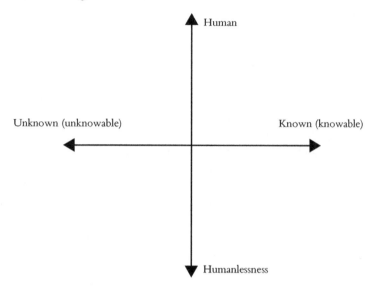

down the diagram a little further. The vertical line represents the competing forces around the inclusion and exclusion of the *human* in algorithmic systems. The tensions occur here over the amount of human agency, as we might put it, integrated into automated systems. This stretches from the pursuit of an advancing humanlessness on one extreme, in which the human might be removed or erased within algorithmic processes, to more defensive and protective responses to automation in which the human is actively reinserted. The former seeks advancement by limiting the human as much as is possible, enabling the technologies space to assert themselves. The latter acts of reinsertion seek to limit human erasure and to re-establish the value of the human and of human decision-making. This reinsertion, as we will see, is not necessarily an outright rejection of algorithms, although it may be in some instances, but it is rather based upon the tension that emerges as the meshing of human and machine agency is both imagined and deployed.

To tackle these issues the chapters in this book work as two pairs. Working across the tensions depicted by the vertical line in Figure 1.1, Chapters 2 and 3 deal with the meshing of human and machine agency. Chapter 2 looks directly at the pursuit of a kind of *posthuman security* in which the aim is to limit human intervention within algorithmic processes so as to manage perceptions of risk. Conversely, Chapter 3 then looks at how the human is placed directly into the active circumscription of agency involved in the navigation of the perceived limits of algorithmic thinking. As such, Chapters 2 and 3 explore the tensions created by the competing moves towards both the integration of the human and an establishment of forms of humanlessness. These two chapters show the competing tendencies that define algorithmic thinking when it comes to the role of the human. These two chapters are suggestive of how these tensions take form.

The horizontal line in Figure 1.1 represents the competing forces around what is known and what is not known. This also leaks into questions around what is knowable and what is unknowable. The book seeks to bring to the surface these elusive tensions around the limits of knowing, suggesting that knowability is a site of tension in which both new knowns and new unknowns are generated. The desire to know through new types of intelligence is tempered by the new forms of unknowing and unknowability that these systems also create. Again, working in a pair, Chapters 4 and 5 deal with the horizontal axis in Figure 1.1. Chapter 4 looks directly at the way that automation is stretching what is known and moving knowledge beyond boundaries. That chapter is concerned with the breaching of boundaries and the stretching of the known. Whereas Chapter 5 looks at the unknown and the unknowability of advancing automated systems. It reflects on how the stretching of the known creates tensions around unknowability. Chapter 5 is concerned with how the unexplainable is actively pursued as part of the advancement of automation. The result of these pursuits creates issues for thinking in straightforward terms about the revelatory potentials of algorithmic systems. In other words, these two chapters deal with the politics of knowing and the way that new insights, knowledge and explanation are in tension with gaps, uncertainties and inexplicabilities.

I should reiterate that I am not dealing here in simple binaries. It is not either the human or humanlessness and nor is it simply either knowing or not knowing. These are forces in tension. It is not about the *either-or*. This book's chapters are about interplay rather than dichotomy. Alongside this, what I am exploring here is not anchored by a fixed notion or definition of the *human* or the *known*. Instead the book reflects on how those notions are defined and redefined in these tensions. I do not seek here to create distinctions or to draw definitive lines; rather the aim is to see the tensions that might implicate the drawing of such lines and facilitate their fluctuations. With this in mind, the chapters that follow don't completely separate out the forces with which they are concerned; instead they narrow the focus to particular specifics while maintaining the tensions with other forces. As each chapter focuses on one set of specifics the other forces and tensions remain in play.

Across the following chapters various resources are used to create insights into these relations, tensions and forces. The case studies and other materials used are intended to be illustrative and to open up the conceptual issues rather than being definitive or providing a complete perspective. These chapters provide glimpses into the depths of the tensions of algorithmic thinking. Indeed, part of what is being argued here is that a response to the very dynamism of algorithmic thinking is required, particularly as the technologies change, social and individual responses to them are redrawn and as those aforementioned myths of a new life are reformulated. The issues that these particular chapters are used to animate are too dynamic and broad to simply be dealt with in complete terms or with just one conceptual angle. These chapters call upon various resources to highlight and flesh out a range of conceptual issues. From this, the focus is upon developing four specific concepts but on the provisional basis that more will be needed. Each chapter uses a set of materials to bring out a conceptual issue or focal points that might be used to explore the broader forces and tensions of which they are a part. In Chapter 2 the concept of *posthuman security* is used to think about humanlessness. In Chapter 3 the notion of *overstepping* is used to think about the reintegration of the human. In Chapter 4, influenced by Katherine Hayles, it is the concept of the *super cognizer* that is used to explore the stretching of the knowable. In Chapter 5 the concept of nonknowledge, reworked from the musings of Georges Bataille, is drawn upon to think about the unknown and the unknowable in automation. Each concept is intended to provide one particular opening upon the broader tensions of which they are a part.

I will not delve any further into the detail now – this will be done within each of the chapters – instead, I simply want to use these opening pages to point to the broader tension-making and tension-holding that defines algorithmic thinking. As I have suggested, these tensions occur both where algorithms are *thought of* and where algorithms are used to *think with*. The suggestion is that these tensions and the competing forces behind them are an integral and implicit part of the various systems of knowing and automation that are continuing to expand. These changes will continue, and probably at some pace, yet my argument is that automation and *algorithmic thinking will always be tense*. The forces and tensions may change form and adapt to the circumstances, and the whole of the space demarcated in Figure 1.1 may be pulled in different directions as one of these forces gains traction and moves the whole picture. Whatever movement occurs, these tensions will remain. Put simply, understanding algorithmic thinking is to understand the tensions that it is shaped by and that it comes to shape.

2

The Pursuit of Posthuman Security

In the summer of 2018 the technology company Hdac ran a television advert depicting their version of the 'smart home'. With clean lines and neutral colours, the automated home space was a picture of hyper-functional minimalism. In many regards it was an entirely unremarkable advert – its style and tone were comparable with typical technology company promotions. Despite the familiar stylistic features, it was the advert's very prominent mentions of 'blockchain' that was particularly notable. For the first portion of the advert a small message appeared at the bottom of the screen telling the viewer that 'Hdac Technology is building the future with the blockchain solution'. This talk of future-building immediately returns us to Lefebvre's impression of the *new life* as discussed in Chapter 1 – the blockchain home is also just around the corner, it would seem. Later in the advert the voiceover reiterates the central message, adding that the 'Hdac Technology platform is smart and secure thanks to the blockchain solution'. This text appeared again at the bottom of the screen. It is clear that the main message of the advert is that blockchain is responsible for enabling the various visions of convenience and technological adaptability being depicted. The blockchain is also given responsibility for ensuring the security of these spaces. Despite its prominence in the messaging, the advert did not go on to say what blockchain was, nor did it mention its functionality or how it was to be applied. Blockchain instead seemed to stand in for a secure (while non-specific) technical apparatus. Blockchain was itself the message. This advert is illustrative of how blockchain is associated with notions of ideal types of data security; it is also illustrative of how the term blockchain can even act as a byword for technical systems that ensure this security.

Given its unexplained insertion in this advert and the fact that it is a term that can be used without the need for definition, it would seem that the concept of *blockchain* has already become a fixture of a wider technological and perhaps even public and media discourse (as discussed in Chow-White et al, 2020). It is a recognizable term. It is an established signifier. Back when this advert aired in 2018 the term seemed to already carry some weight and recognition. The minimalist confidence in which it is incorporated as the key term in the advert is suggestive of this recognition. As I will explore in this chapter, this confidence in the security of blockchain can be understood in broader terms as operating as one instance in which algorithmic technologies are presented as an antidote to a wider sense of vulnerability and precarity. This sense of vulnerability is often associated with the perceived potential unreliability of human interventions within data processes. As I will explore in this chapter, this type of human insecurity is appeased through the pursuit of various kinds of *posthuman securities*. Blockchain provides a focal point for exploring one prominent type of posthuman security. As this chapter will describe, this type of algorithmic thinking is based upon a vision of what can be achieved through humanlessness or by reducing the level of the role of the human actor in the meshing of various types of agency. Although, this, of course, does not mean that questions are not also raised about trust in the automated aspects of blockchain or with their underlying algorithms (see Zou, 2020: 657). These are part of what I am attempting to articulate here. Indeed, a focus on questions of security brings out the types of tensions of algorithmic thinking with which this book is concerned. Taking blockchain as its focal point, this chapter reflects on a particular instance in which the absence of the human is embedded into visions of a secure society. These are tense securities.

Tense securities

Blockchain has been gaining a great deal attention over recent years, especially as organizations seek options for storing and retrieving the vast data that are now accumulating. The value placed upon data by different industries and institutions means that the safe and reliable storage of those data is now considered to be of paramount importance. This imperative of secure data storage and retrieval goes back through the history of data warehousing but has now escalated with the increasing value attached to data (see Beer, 2019a: 120–6). Taking two illustrative focal points – the art market and the smart home – this chapter looks at the promises and potentials that are attached to blockchain along with the data anxieties that it is said to resolve. The chapter looks at the type of untainted data traces that blockchain's distributed systems are said to afford. To do this, the chapter shows how these visions combine into a dream of a kind of secure society – which is dealt with here as an indicative example of the kind of myths of an algorithmic new life discussed in Chapter 1. It is argued that at the centre of this set of dreams is a form of posthuman security. Posthuman securities, as I will argue, can be understood to be a reaction to perceived human insecurities and vulnerabilities; they are a way of reducing the involvement of the human so as to create a perception of ensuring something more secure. The chapter argues that engaging with the tensions between human insecurity and posthuman security is crucial to understanding the ongoing implementation of blockchain and other related algorithmic and automated technologies. In other words, the posthuman securities identified here operate in various spaces in which the human is sought to be bypassed in order to minimize senses of risk and vulnerability. This is a thread that runs through many types of automation; blockchain is useful in taking us through a particular aspect of humanlessness and the tensions of algorithmic thinking that might be associated with it.

Returning again to the example Hdac for a moment, the various mentions in the advert suggest that blockchain as a technical term has, like 'algorithm' and 'data' before it, already become an established part of a broader vocabulary. It has not gone mainstream to the same extent as those other two terms, but it is perhaps on the cusp of doing so, especially because of the type of just-around-the-corner new life that is associated with it. Equally, its popularity may die away if anything occurs to dent its image or if another technology or term comes along that usurps it – even if this happens I'm hoping that the conceptual ideas used in this chapter will remain, particularly as the type of tensions discussed here are likely to define whatever comes next. Like data and algorithm, blockchain is a technological term that is rarely fully explained in advertising, promotional materials and the like, if indeed a single explanation is possible, yet it is widely used to evoke certain traits, properties and connotations. Blockchain would seem to have quickly achieved a level of notoriety that means that it needs no explanation. It is what the term stands for rather than its exact features that seems to matter.

Blockchain as a term may not necessarily be fully explained in order for it to evoke ideas of a cutting-edge technology, the advancement of computing and, crucially, for it to be uttered in order to promote the impression of an infrastructure that can be trusted. Or, more specifically, the term blockchain can itself infer a system that can 'eliminate the need for trust between parties', even where they might not know each other or have contact with one another (De Filippi and Hassan, 2016). The trust placed in the blockchain technology enables 'trustless' (De Filippi and Hassan, 2016) exchanges between 'untrusting participants' (Chow-White et al, 2020: 3). In this it is clear that trust is complicated by these systems, leading to suggestions that this is the 'emergence of trust in a new form' (Zou, 2020: 646). In this regard, the Hdac advert captures something that this chapter seeks to explore further: the wrapping up of blockchain within dreams of a secure society and the potentially tense relations between algorithms and humans on one hand and trust and security on the other. Perhaps, as Robert Herian (2018: 170) has argued, the development of blockchain might tell us something about wider developments in 'capitalism', 'reason' and 'data' (a similar point is elaborated with a narrower focus on Bitcoin by Swartz, 2018). Blockchain is an active presence in these processes. Indeed, Werbach (2018: 4) has argued that 'the soil in which blockchain took hold was the crumbling trust in governments and corporations amid the financial crisis of 2008 … the erosion of trust endured'. It was the conditions of an eroding trust that was the backdrop to which the ideals of trustworthiness were attached to blockchain, this collapsing trust creating a space for its expansion. It also meant that human vulnerability and insecurity were heightened.

One specific promise attached to blockchain is that it enables the untainted storage of data. It is presented as the means to a kind of pure or perfect archive. As I will explore in this chapter, blockchain is understood to be made up of documentary traces that cannot be changed or tampered with or 'retroactively modified' (De Filippi and Hassan, 2016). This unalterability is due to its diffuse and distributed traces. I am emphasizing this point before elaborating on the details because it immediately points towards the building of trust within distributed systems and a concern, more specifically, with what has been described as blockchain's 'architectures of trust' (Werbach, 2018). I'm not focused in this chapter on whether technologies like blockchain actually achieve the securities that they are claiming to achieve; instead I am concerned with how trust takes shape, the tensions this version of trust is forged from and how this trust is actively built into and projected onto the systems themselves. As Nelms et al (2018: 24) have observed in the case of Bitcoin users, 'trust is established by code'. If this is the case, then how that trust is coded into a system is worth attempting to unpick.

Much of the work on blockchain has understandably focused upon its use in the development and operation of cryptocurrencies such as Bitcoin (see, for instance, Bjerg, 2016; Nelms et al, 2018; Hayes, 2019). Understanding blockchain in the context of currencies is crucial and so are the writings on this topic, but I'd like to think more broadly here about how blockchain is being pursued elsewhere and in different settings. As I have mentioned, two alternative focal points have been drawn upon in this chapter: the art market and the home space. These have been chosen because they are spaces in which security is considered paramount, even if in quite different forms. They are also useful because they allow us to extend the exploration of blockchain's role and to reflect further on the types of tensions that emerge from the algorithmic thinking as exercised in relation to blockchain. In each case the chapter will look at how trust is being built and how possibilities and opportunities are being envisioned through the promotion of different forms of humanlessness. In short, the chapter looks at a set of moments where technological and automated solutions are sought to address the unease and precarity of data-led social ordering. Let us look first at blockchain's promises of untainted knowability.

Untainted knowability in the blockchain

The sense of security provided by blockchain is built around the automated distribution of records beyond direct human involvement. It is pictured as affording a kind of unhuman archive, populated and managed by 'algorithmic archons' (Beer, 2021). To develop some of these sentiments further, let us dip into a piece of journalism on blockchain's worth. This illustrative passage brings to the fore some relatively widely expressed views on blockchain:

> bitcoin might be newsworthy, the really important story concerns the blockchain technology that underpins it. ... Blockchain technology is indeed important, but it seems largely incomprehensible to ordinary mortals, even though the web teems with attempts to explain it. ... But implicit in the blockchain concept is the endearing strain of technocratic utopianism, a hope that technology can overcome some aspects of human frailty and corruption. (Naughton, 2018)

As with the example from the Hdac smart home, this passage indicates that blockchain has moved beyond specialist terminology. Despite this, it somehow remains mysterious in its technical details. It is obfuscated, somehow, by its technical complexities. This passage also fits into the findings of Chow-White et al (2020) who noted that in media coverage blockchain is more often seen as a positive innovation whereas Bitcoin tends to be associated with more sceptical feelings. This one instance is illustrative of the frequency with which blockchain is evoked to suggest the possibility for enhanced security, and how blockchain in general is seen as safer than its specific application in Bitcoin. Behind this is the notion that blockchain is secure because it is able to bypass human intervention. Crucially, blockchain is a technology that appears to, as it is put in the previous passage, 'overcome human frailty'.

In an introduction to this technology, Malcolm Campbell-Verduyn (2018: 1) writes that 'at their essence, blockchains are digital sequences of numbers coded into computer software that permit the secure exchange, recording, and broadcasting of transactions between individual users operating anywhere in the world with Internet access'. And so here we see in this introduction the central idea of the facilitation of secure remote transactions – giving a hint also at how security is built into the rationalities of blockchain. In more material terms, blockchains have been defined as 'distributed computing technologies that securely record data on append-only digital ledgers and execute code' (DuPont, 2019: 29). It is vital to note that these ledgers are decentralized. Again security is foregrounded in the description. The notion of append-only records and the decentralized nature of the storage are crucial starting points for understanding blockchain as a concept.

Emerging out of the work conducted for the establishment of the Bitcoin cryptocurrency, 'in late 2013, the development of what was by this point called "blockchains" (as one word) started to take off' (DuPont, 2019: 84). This then is a relatively recent term that has risen quickly. In essence, blockchain is a way of recording data across distributed computer networks. As has been suggested elsewhere:

> The term blockchain has gained currency in recent years as the name for a particular type of distributed system, popularised by the cryptocurrency Bitcoin. In technical terms, a blockchain is simply an append-only data structure, consisting of a chain of data blocks linked together by cryptographic hashes. Such chains were first developed in the 1990s as a data integrity technology, to prevent the tampering of records held by an organisation. They were used for this purpose long before Bitcoin gained prominence. (MacDonald-Korth et al, 2018: 7)

Dating back to around the 1990s, we can already see the perception of a need to prevent alteration, which has been blockchain's central feature from its inception. The idea that a record can only be appended or added to rather than changed is important in the conceptualization of blockchain and its possibilities – it is also a common and often-repeated feature of definitions of blockchain (there are many video explainers that are now openly available to access). There is a close association between blockchain and cryptocurrencies, especially Bitcoin and more recently related developments such as 'stablecoins' (Murphy, 2018). Beyond this though, and despite the close intertwining of their development, the applications of blockchain go beyond this particular sector. Blockchain technologies are relatively recent but can be placed in a line of development that includes 'chain'-type data-integrity technologies from the last two or three decades. So there is a longer history to the type of technology that blockchain is associated with, while that particular term and some of the properties associated with it are more recent. Indeed, despite these types of technology having been established some time ago, the term blockchain has only really been used widely in the last decade, and particularly in the last few years. With the reliance on data so too come concerns over data integrity (see Beer, 2019a). The move towards a wider consciousness of the term has been even more recent.

Feeding back, there were specific developments around Bitcoin that enabled the technological developments of blockchain. The rise of Bitcoin and the extent of its coverage also then had the effect of pushing blockchain technologies into the popular imagination. The history of the relationship between blockchain and Bitcoin is important for understanding both. The developments fed between Bitcoin and the wider development of blockchain. Even if we are to look at other applications of blockchain, Bitcoin is still part of the story of its development. According to MacDonald-Korth et al (2018: 7), 'Bitcoin's novel innovation is the so-called "proof-of-work" algorithm'. It is this 'Proof-of-work [that] makes it possible for a chain of data blocks to be maintained on an open network that any computer can join, without compromising the integrity of the data' (MacDonald-Korth et al, 2018: 7). It was this ability to manage a chain of blocks across a distributed network that is identified as being the crucial step here and it is also where algorithms play a part in managing that distribution and recording process. The move that is being identified here as being crucial is the algorithmic management of a diffuse distributed network of data storage. The integrity of the network is understood to be maintained through this automated decentralization – which in turn creates questions about security at what then appear to be the vulnerable 'edges' of computational networks (see for example Swabey, 2022).

The result, it has been suggested by Quinn DuPont, is that there are two types of mediation going on with blockchain. First, DuPont (2019: 82) argues, 'blockchain technologies are media in that they sit between people, carry meaning, and transform messages'. This first form of mediation is the mediation of the social networks within which transactions are produced. Whereas, second, 'blockchain technologies draw on their arrangement of cryptographic algorithms to abstract away the complexity of objects and things, which makes the objects and things easily managed' (DuPont, 2019: 82). As well as the relations being mediated, the blockchain is also mediating the objects that it abstracts into data. The first mediation is about the passing of information between people or systems; the second is about the turning of objects into data. DuPont is attempting to think here about the space that blockchains can occupy and how they might then mediate particular sets of relations and connections. The double mediation being described here is an important part of the material relations that blockchain intervenes in and, as we will see, might come to reshape.

In terms of structure, DuPoint observes that 'at its most basic level there are three key components to a blockchain: hashes, blocks, and chains' (DuPont, 2019: 85). These three features are often the focal point for explanations of the structures of blockchain, with each of the three components performing a particular role in the storage processes these technologies facilitate (for a further example of the description of blockchain through a focus on hashes, blocks and chains see Lim, 2016). Hashes are the identifying marks within blockchains. The hash is created by a 'hash algorithm'. The hash is central to the other two features of the blockchain, because 'hashes are used to create and identify blocks of transactions and then chain them together' (DuPont, 2019: 85). Before providing a detailed analysis of the SHA-256 algorithm used in Bitcoin, Donald Mackenzie (2019: 35) explains that a 'hashing algorithm takes a message, scrambles it thoroughly, and condenses it into a relatively short, fixed length form called a "digest"'. The hash, then, is a kind of marker. According to DuPont, 'the hash algorithm creates a unique identifier (or digital "fingerprint") for transaction data' (DuPont, 2019: 85). Produced by an algorithm, the hash here is compared to a fingerprint for the transactions that are being recorded. This fingerprint is then fixed – fixed by the algorithm. The algorithms that produce them are unidirectional and the outputs can therefore be checked but are very hard to reverse. DuPont (2019: 86) explains that 'Hash algorithms are deterministic, so identical inputs will always produce the same outputs. … This means others can check the validity of a hash by re-running the algorithm on the original input – if the test matches the expected output, then the hash is valid' (DuPont, 2019: 86). So results can be checked and cross-checked to see if the hash is 'valid' and if the blockchain is untainted. If the input is the same the test will always produce the same output. This is the basis, then, of a type of 'algorithmic authority' in blockchain (Lustig and Nardi, 2015; also discussed in DuPont, 2019).

The second feature of blockchain is the blocks themselves. Essentially, a block can be understood as 'a list of transactions' (DuPont, 2019: 87). It is by connecting together transactions that blocks are created. This makes the connection of transactions possible and therefore is central to the formation of blocks. DuPont explains that 'hashing transaction data produces a uniquely identifiable block' (DuPont, 2019: 88). Each block is unique and is marked with the unique hash. In short, the 'hash value serves as a helpful identifier for the block' and, at the same time, 'the hash ensures that no transactions have been altered after the fact' (DuPont, 2019: 88). These blocks have a unique hash that means they can be identified and verified, and so can their place within the chain. Taking Bitcoin as an example, MacKenzie (2019: 36) also notes that:

> When bitcoin miners have the current block of transactions, they also incorporate the hash of the previous block, which in its turn includes the hash that came before it, and so on all the way back in time to Satoshi's 'genesis block'. Suppose just one aspect of a single transaction is altered (perhaps several years ago someone received one bitcoin and now tries to alter that to ten bitcoins). It isn't just the hash of the old block that would completely change. The hash of every subsequent block would too, making it clear that the blockchain had been tampered with.

It is this linking together of blocks into a chain that leads back to an original transaction that is the important image here. If one change is made it becomes visible throughout that chain. The result is not just unalterability but the sense that any changing or tampering is highly visible and can be tracked (which of course then produces questions about the surveillance made possible through such visibility, as discussed by Inês Faria, 2019: 129). This then is not just an archive that captures data traces but that also automatically tracks any changes to those data traces – traces of traces.

The block's unique hash, as mentioned, also means that the record of transactions can be checked and verified. At this point, 'once a hash of transactions is created, the block can be independently verified to prove that none of the transactions have been altered' (DuPont, 2019: 88). Verification and unalterability are key features of the blockchain. If any changes have been made to a block, the input would be different and so the hash that identifies the block would have altered (as discussed by Marshall, 2018). Alongside this, further reassurance is sought in the very structure of the network that supports the production of the blockchain. The use of a 'distributed computer network' means that, through diffuse connections and the involvement of multiple devices or nodes 'no single computer on the network is essential for the network's overall performance' (DuPont, 2019: 89). As I will explore, this means that there is understood to be no one single point of vulnerability. We see here further attempts at layering security and trust into the structures of the blockchain. It also means that multiple devices and actors are involved in verification – which is suggestive of an increasing of the accuracy of the records that are captured. It is for this reason that there is understood to be no necessity for trust between parties involved in transactions. DuPont explains that this is a form of 'distributed validation process' in which 'each miner independently verifies that transactions are well formed and comply with a specified set of criteria' (DuPont, 2019: 91).

The third and final feature that DuPont identifies is chains. These are produced as the hashed blocks are connected together, with each block containing the previous hash. DuPont notes that 'just like transactions, blocks are chained together for security' (DuPont, 2019: 96). So, transactions are connected together to make blocks and then blocks are linked together to form chains. These chains are crucial to the functioning of blockchain. The connections in the chain are what make the record trackable. The hash is crucial in the linking of transactions into blocks and then in the linking of blocks into chains. It is, DuPont explains, 'by including the previous block's proof-of-work hash, a long chain of hashes is created. If any previous block's transactions are altered the proof-of-work hash will be invalidated, and any block with a new (altered) hash will cause a ripple effect up the chain' (DuPont, 2019: 96; see also MacKenzie, 2019: 38). The connections or linkages in the chain are formed through the inclusion of the previous block's unique identifier within the next. Changes create ripples up the chain. The idea here is that the chain can then be checked and verified to ensure that no changes have been made to any of its components, with the automated hash and the storage across devices securing the blockchain. It is this irreversibility, unidirectionality and identification that is crucial to the claims to security made by those deploying blockchains. Organizations are overhauled in the move, DuPont (2019: 193) surmises, towards 'decentralized autonomous organizations'. And so, to explore this form of algorithmic thinking in more detail, let us turn to these decentralizations and to the roll-out of blockchain across two particular sectors. I'll start with the art market.

Perfect provenance? Blockchain in the art market

If there is one issue that is of paramount importance in the art world it is provenance. Perhaps it was inevitable that blockchain would be offered as a solution to ensuring the establishment and accurate tracking of provenance. As it has been claimed, it is here that blockchain has been envisioned as having the 'potential to improve provenance, records of ownership and proof of authenticity of an artwork' (Bailey, 2018). The claim that is often made is that blockchain might reduce if not eradicate the impact of forgery and theft. Threats and dangers are the backdrop to the formulation of the promises of security that are attached to blockchain. Trust is built in the technology on the back of a foundation of perceived risks. This has been thought of in terms of the removal of risk and even in terms of notions of 'de-risking' (Faria, 2019). As it is pictured, the potential of blockchain is to placate these all-too-human risks. So, for instance, it is observed that:

> Art is currently plagued by fraud, illicit business, and tax evasion, all products of a fragmented physical market that is hard to follow. Enter blockchain, which on the surface appears a silver bullet. In one shot, blockchain could ensure the veracity of an art piece, make the price and parties to a sale transparent, and allow oversight to monitor the flow of art assets in and out of different tax jurisdictions. (MacDonald-Korth et al, 2018: 5)

The authors of this report are observing a general sense of hype and of the type of potentials that are associated with blockchain. Crucially, it is the fear of the art market being undermined that is a driver. That is to say that a fear of the vulnerabilities within the market are prominent in creating the space for the implementation of blockchain. The authors are reporting on the type of optimism that is prevalent. Following this passage the authors go on to acknowledge the clear obstacles for blockchain. What they identify here is the idea that blockchain is able to resolve these problems. And so we can observe that a tangible sense of vulnerability within the sector is part of the move towards blockchain. Blockchain is presented as the means by which those vulnerabilities might be curbed.

Fears of fraud and other problems permeate through the discussion of blockchain's potential. It has been observed, for instance, that the art world is experimenting with blockchain and is 'seeing technology as a way to revolutionize how art is bought and sold, thwart fraud and tax evasion, and reduce friction during the auction process' (Gagne, 2018). The concerns are both with being misled over the history of an artwork, with losing records and with the financial losses incurred through inefficient market processes. In this context, in which the vulnerabilities of the market are long-held but are being more acutely felt, blockchain is presented as a direct solution to those vulnerabilities. The algorithmic new life here is based in a sense of security and efficiency that resides just around the corner (as discussed in Chapter 1). A kind of dream-like near future is then imagined in which blockchain shores up the art market and produces a perfect market square with trustworthy trading.

We can see this relation between fear and a more secure future playing out across the blockchain discourse. Art Market Guru (2019), a website that tracks changes in the art market, acknowledges that:

> Blockchain is the new Tech buzzword in the art market. Every month, it seems, a press release comes out announcing yet another 'disruptive' blockchain application. The claims include a range of industry-shattering functionality, including: the end of issues involving provenance, the ability of artists to collect commissions across the lifetime of a work, the end to fraudulent works placed on the market, and the democratization of art.

There is then an awareness in the sector that blockchain is a buzzword and that there is a significant build-up of hype around the technology. There is also an awareness that it may not achieve the mythical new life of a perfect security that is often envisioned, yet the pull of blockchain's promises seem to remain powerful. A perpetual stream of idealized innovation is identified in the previous passage, with blockchain a panacea that can adapt and be utilized to address different shortcomings in the structures, approaches and records of the existing art market. Even within this discourse it would seem there is a tone of weariness and wariness being expressed. Rather than rejection, the approach seems to be to try to sift out the genuine. However, the kind of adherence to the promises and dreams of blockchain seem to be remarkably durable, even where this slight air of scepticism, and it is often only slight, can be detected. The crucial dream in the art market that holds firm is that of the achievement of perfect provenance.

As with the wider trust in blockchain, it is the process of the distribution of traces across networked devices that is seen to present possibilities for solving the problems of the art market (MacDonald-Korth et al, 2018: 13). In the art market, where trust is important, the potential benefits of untainted records carry particular weight. Trust in what an artwork is said to be and where it is said to originate from are paramount to the functioning of the art market and to the measures of value within it (MacDonald-Korth et al, 2018: 13). Blockchain is imagined to provide potential opportunities for a more trustworthy and less human-orientated or human-reliant tracking of provenance and ownership. Heightened senses of vulnerability of record keeping present an opportunity for those seeking to expand the reach of blockchain. And so here is an instance in which the reduction of the role of the human is a central part of these forms of algorithmic thinking; this inevitably creates tensions.

On the surface, centralization would be the most likely feature of trust building, but in the art market it is the potential for decentralization that seems to be fuelling the drive towards blockchain (for example, see the descriptions in Mire, 2018; for a further example see Bailey, 2018). Again, trust and distribution go together in these visions, especially where 'distributed ledgers could be used to both track the ownership history of a given piece, and prove the provenance of the piece simultaneously' (MacDonald-Korth et al, 2018: 13). In their report on the art market, MacDonald-Korth et al (2018: 14) have pointed out the complexity around transparency within the relations of the art market. They note that for a variety of reasons some are likely to prefer secrecy whereas others may prefer transparency (as discussed by Sprague and Cameron in Art Market Guru, 2018). Despite any potential cultural barriers, the discussion is often of how 'the distributed ledger known as blockchain' can be used to 'bring transparency to their operations' (Gagne, 2018). Zohar Elhanani, a CEO of an art market data analytics provider, observes that blockchain is 'being introduced to broaden the market's transparency, track ownership and provenance, and provide an infrastructure for the tokenization of fractional artwork sales' (Elhanani, 2018). Blockchain and ideals of transparency are closely related (and I'll return to the point about fractional sales in a moment). As such, it is also, as these accounts of the art market suggest, part of the wider technologically fuelled 'politics of opacity and openness' described by Clare Birchall (2011). And so trust is built through the ability to make transparent by decentralizing and moving the storage of data away from human intervention.

More specifically, the accounts and speculations in this field suggest three main applications for achieving this combination of transparency and security. First is tokenization. Tokenization facilitates the trading of portions of an artwork (see Elhanani, 2018). An artwork is effectively turned into tokens or fractions that can be bought and sold. The parts of an artwork can be owned based on trust in the blockchain and can be held by people who have no connection to each other (this is the trustless relation described earlier in the chapter). This tokenization occurs often where 'portions of blue-chip masterpieces are traded like assets' (Shaw, 2018). And so the painting can effectively be traded in pieces rather than as a whole. Clearly this opens up that particular high-value artwork to a different market and to different buyers. It is also suggested that tokenization allows the artist to more easily sell parts of work as a 'token on the blockchain' (Gagne, 2018). The conclusion is that this also makes it easier to sell lower-value works too. The blockchain in these instances allows the artwork to be broken apart into smaller transactions that can be individually tracked to manage the overall ownership and flow of revenue. Aerum (2019), a provider of such blockchain tokenization, claims that tokenization through blockchain provides 'solutions' for 'removing friction, middlemen and gatekeepers'. This notion of increased efficiency and the loss of friction often appears in accounts of blockchain, as does the proposed removal of 'middlemen' and 'gatekeepers'. As we will see, the bypassing of these figures represents the removal of human vulnerabilities. These figures become the embodiment of those senses of vulnerability. As it was put in a separate instance, 'the machine acts as the middleman' and so blockchain can 'eliminate the need for middlemen' (McClintock, 2017). It seems that there are concerns about the vulnerabilities and inefficiencies caused by intermediaries in such transactions (see Werbach, 2018: 78). The image that is created is of a more direct set of transactions that go unmediated other than by the automated blockchain ledger. The old ideals of disintermediation that have long circulated in web technologies have found a new type of presence and pace in blockchain, with the apparent and widely discussed potential for 'radically disintermediating institutions' (Hayes, 2019: 66). The very notion of disintermediation is interesting here, as it is suggestive of how the social relations of the art market might be recast or even usurped by blockchain.

Second is the shift in the recording of transactions. The idea here is that blockchain can be used to record transactions in ways that enable them to be tracked and recovered. Nothing, according to such premises, can be lost. A notion of a pure archive could be another way of understanding what is envisioned. The ideals of the perfect recoverability of transactions and the untainted knowability of the decentralized archive is again found here. Christie's auction house, for instance, are reported to have 'partnered with blockchain-secured registry Artory'. The outcome of this is that some transactions are now recorded entirely through a blockchain (Elhanani, 2018; see also Gagne, 2018). The result is the claim that blockchain can be applied to achieve a comprehensive and unalterable record of transactions. In this instance, 'Artory's registry tracks histories, provenance, and archival material while allowing buyers to remain anonymous, increasing buyer and seller confidence' (Elhanani, 2018). Not only does the use of blockchain increase the reliability of the record of transactions, it is suggested, we also see here, how it is presented as enabling anonymous transactions to occur. This illustrates how these seemingly incompatible ideals of recordability and anonymity mix together. The blockchain, it would seem, is both more transparent and more opaque.

Alongside these, and closely related to them, the third move is to using blockchain as part of the implementation of digital-only transactions, which is already occurring in some blockchain-based auction houses (see Elhanani, 2018). This is simply about recording all art transactions within a blockchain as part of a wider move to only facilitating digital transactions. The move here is quite a basic one, but it facilitates the other changes and suggests that it is possible for the art market to move transaction records entirely onto blockchains. In this sense, this is a foundational step which is already underway. Such examples of blockchain-based auction houses become suggestive of a vision in which archiving becomes entirely automated and decentralized.

The changes and possibilities inevitably create questions over the future of the art market's intermediaries (Shaw, 2018). Here the tensions of humanlessness arise within such algorithmic thinking. As we have already seen, this removal of intermediaries in the development of the art market is an often-stated aim in the implementation of blockchain. To highlight another instance, it is noted that the AllPublicArt platform adopts blockchain technology to allow 'art collectors to do business with each other through smart contracts, eliminating the need to pay commissions to intermediaries' (Art Market Guru, 2019). The impression here is of the removal of unnecessary layers and that a more direct connection is facilitated through this seeming building of trust in blockchain. This removal of the intermediary is frequently presented as being the means by which improved transactions and more secure tracking of the records of the art market can be achieved. The human intermediary is presented as the weak link when it comes to ensuring the security and reliability of the data. The human intermediary is also presented as a source of inefficiency within these existing processes. In short, the intermediary becomes the target that blockchain is designed to eliminate. There is a sense that blockchain will change the intermediation of art and alter the actors, systems and processes involved. There is also a sense of security being fostered in the non-human aspects of these new approaches. The future art market as it is imagined would be less about human judgement and records and more about the integration of these automated blockchain systems. Indeed, we see here how this is described in terms of technologies that are already in place and that the art market is on the cusp of further changes. This is an image of a kind of trading at a distance, with that distance secured by the apparent infallibility of the blockchain. Inevitably these moves produce tensions and resistance. Elhanani has suggested that, 'as it stands, when buying art, the security of a face-to-face encounter with a specialist is hard to replace, especially with blockchain technology still in its infancy' (Elhanani, 2018; I will discuss this kind of establishment of the limits of automation in Chapter 3). There are different types of security at play here and there are tensions too between human security on one side and technological or posthuman security on the other.

When it comes to both the removal of intermediaries and the tracking of provenance, the establishment of authenticity is a driving presence. For instance, echoing the types of perceived vulnerabilities already discussed, two founders of a company providing blockchain to the art market have claimed that:

> Our initial thoughts stemmed from a concern over the authenticity of artwork and avoiding forgeries, which led us to thinking about the ecosystem of art. When we compare art to other investments or valuable assets, it is noticeable that there isn't a registration process available to record the authenticity of works, especially when compared to cars, property, shares or other securities. Without registration and proof of authenticity there is a large challenge to gain confidence of investors to purchase art particularly from emerging artists. (Sprague and Cameron in Art Market Guru, 2018)

In this statement authenticity is placed as a primary concern. The specific need that blockchain is adapted to solving is the preservation and protection of that authenticity. These founders also specify the importance of confidence in that authenticity. The fear is of the loss of confidence in what is authentic, with an impact then on the value of the asset. From this perspective, the blockchain is presented as a means for gaining confidence in the origin and authenticity of the artwork. And so blockchain becomes somehow a guarantee for the trustworthiness of the artwork and its source. Although, as De Filippi and Hassan (2016) have pointed out, this process can create additional complex issues around 'trustlessness' and the checking of whether the records in the blockchain match with actual exchanges and the like. Yet it remains the case that a significant part of the functioning of blockchain here is in the promotion or establishment of a new-found confidence in the provenance of art and the means by which human gatekeepers can be bypassed. It is often the human vulnerability of record keeping and the potential for gaps, mistakes and misleading records – a human insecurity – that is establishing a path towards blockchain and its promise of a kind of posthuman security. The dream of perfect provenance is also a dream of unquestionable legitimacy and unbendable verification.

Pure connectivity? Blockchain in the home

Such themes go beyond the art market. I opened this chapter with Hdac's use of the term blockchain to reinforce the image of the secure smart home. Hdac provided an example of the defensive fortification of the data-informed home space and, in this regard, an instance of what Rowland Atkinson and Sarah Blandy (2016) have called the 'domestic fortress'. The applications of blockchain frequently fit within that broader picture. Blockchain is offered as the means to fortify the data-intensive home space. Where the home requires data to function it is thought to become vulnerable to data breaches. Blockchain is presented as allowing the home to know its occupants better while also ensuring that that data-informed knowing is protected. The increase in data compels the increase in data security.

As the home becomes more connected so new vulnerabilities are created. Again, such fears create a space for blockchain to expand. New data vulnerabilities come to require a solution. The linkages and interfaces, the points of contact between systems and so on, become the focal point for security concerns. Looking across the moves to implement blockchain in the home, versions of this question of data security occur: 'how' it is often asked, 'can blockchain improve the security of connected devices in smart homes?' (Lashuk, 2020). There are many variants on this question asked by those considering or seeking to bring about a particular future for the so-called smart home. These visions vary somewhat in scale, in some cases focusing on the small objects within the home and in others dealing with the general direction that the home might take in its form and structure. Occasionally these scales combine. One response takes the line that:

> Blockchain improves security and rationalizes operations in smart homes. It powers up transactions in the sharing economy and establishes a secure environment for the remote control over connected devices. It prevents hacks and data breaches, all to make our daily lives more convenient. (Lashuk, 2020)

The image of efficiency and of the remote-control of the home is often combined with that question of security – this is clearly not dissimilar to the combination of themes found in the art market. This sits alongside the sense of a need to ensure that the many connected objects of the smart home both speak to one another and that the network that they form is robust and leakproof. The future secure home network is pictured with the idea that 'blockchain will keep proving itself useful across your connected lamps, blinds, smart locks, TV sets, other home appliances' (Lashuk, 2020). The mundane objects of the home, including appliances such as the lamp or the TV, become objects to be connected and then secured. It is this set of outside connections to the networked objects of the home space that create one focal point for provoking a sense of vulnerability. It is such connections between the home and broader networks that are presented as the point of connection where the additional security is thought to be needed. The interfacing of the private home into what are seen to be non-private networks is, it would seem, potentially unsettling. It is a move that creates an apparent discomfort or fear. The objects of the home then become something to be protected or secured. As well as outward-facing interfaces, the inward-facing connections are also foci. This type of device-focused description in this passage in which ordinary objects – such as lamps, locks and TVs – are networked together can be found in various places. It is suggested, for example, that blockchain can be used 'for the secure management of multiple smart appliances, whether that's an air conditioner or juicer' (Jones, 2018). Suddenly, in this home space, even the juicer is something to be controlled and secured. The blockchain is understood as the source of future security of these object and spaces. It is also associated with managing the various connectivities, networking and control of the home and of the ordinary objects within it. While doing this, there is also a repetition of a theme from the discussions of the art market, with blockchain associated with increasing efficiency.

A prominence of a sense of control is important in understanding accounts of blockchain in the home. In another instance, in a dedicated video on the use of blockchain within smart homes, the technology company IoTex promotes its use of blockchain in terms of the expanded controllability of the home (IoTex, 2018). They envision 'how homeowners can control and share their smart homes'. In this case, the point is to offer short-term home rentals using a decentralized blockchain (which is perhaps a similar type of application to the tokenization of artworks discussed previously). This particular service adds the 'control' of the smart home through app-based remote monitoring of 'smart locks, thermostats and lights'. It adds to this the control over access to the rental space – a type of control over the controlled space – with rentals and access to the home controlled through the blockchain. This captures the image of heightened control that is associated with this implementation of blockchain, and emphasizes how this control can occur at a distance while still, as the underlying theme repeats, in a secure form. With, in this instance, blockchain said to enable users to 'securely authorise access to their smart home' (IoTex, 2018). A sense of security through proximity is extended outwards to create security at a distance.

As we have seen in those connected but ordinary objects, when applied to the home there is an obvious and inevitable connection between blockchain and what is often referred to in popular internet culture discourse as the internet of things. As objects, spaces and bodies are networked in these home spaces, so the data gathered inevitably tracks these domestic 'things'. Once networked, these items of private domesticity call to be made secure.

On this point, the blockchain service provider Openledger (2020) claims that: 'Blockchain makes Internet of Things secure, transparent, and tamper-proof.' Of course, no one would want their connected and smart home space to be hacked. Such a hack might be associated with the loss of control over the data and a loss of control of the networked objects and spaces within the home. Clearly there are echoes here of the broader appeal of the untainted traces of blockchain as discussed earlier in this chapter. As is often the case, it is the automated nature of these traces that is central to the authentication and legitimacy of the information being held. It is this very movement of data records beyond human intervention that is important in understanding the sense of security and control that is projected onto blockchain. Echoing the broader ideals of the 'permanent record' (Campbell-Verduyn, 2018: 1), this type of security is encapsulated in Openledger's further claim that 'blockchain automates security and auditability, delivering verifiable, secure and permanent data storage'. The picture builds of the focus on security that is associated with blockchain. It is the secure connection of objects that is placed at the centre of these accounts.

What is striking is just how consistent the framing of the blockchain-based home is. Indeed, the dream of a predictive or knowing home, which William J. Mitchell (2005) outlined and explored some years ago, is remarkably consistent in the blockchain-centred form it now takes today. Working through the coverage of the sector, it is not long before a kind of saturation is reached in which the same or very similar points about blockchain are being made repeatedly. Another blockchain service provider, for instance, makes similar claims to those mentioned previously and emphasizes a number of the same points about security and control, as well as highlighting similar fears to which these features are addressed:

> Our Blockchain-enabled Smart home affords privileged Blockchain solution to the owners for controlling or tracking their smart home to get rid of hazards in terms of providing impeccable access to a smart device, immutable, high security and enable other parties for accessing specific areas without allowing them for full access. (Osiz Technologies, 2020)

Here again, the central thrust of this statement centres on the removal of hazards – human hazards it should be noted – and the establishment of control, trackability, complete connectivity and, of course, the security of the data and of the networked components of the home.

To reiterate, as the discussion so far might suggest, it is the connectivity of the components in the smart home that is perceived to create new security issues. There is a sense of fear of the new permeability of the home that is generated by its even more highly 'networked logjects' (Dodge and Kitchin,

2009: 1351). The networked vulnerabilities of the smart home create certain concerns and data anxieties. These are evident in the following sentiments:

> The concept of the smart home has been around for many decades, but it is only in recent years with the advent of the so-called 'internet of things,' IoT, that meters and monitoring, cameras, locking systems, heating systems, and entertainment and information devices, have led to many homes having some degree of genuine smartness. Of course, with connectivity and utility come security problems. For instance, a malicious third party might find access to the home's wireless network, reprogram the smart TV, turn up the heating, disable the air conditioning, or even open the front door and allow them to remove all your smart devices and redeploy them elsewhere. (Bradley, 2019)

This is the fear of the out-of-control home in which control is lost over the domestic space. Clearly as the home is networked it generates new senses of vulnerability and insecurity. This passage encapsulates such fears of the home being hacked and control over it being lost as a result. In this gap and in response to such fears blockchain is presented as both *the means of connection* and *the means of protection*. It is promised to provide both of these things. It is understood to allow connections while also shielding those connections from risk. The expansion of data-led devices can then be seen to continue with blockchain providing this kind of antidote to anxieties over data control. The implication is that the potential insecurities of the smart home can only be resolved through the intervention of blockchain – which is essentially presented as allaying the very fears it provokes. Or, as it is put in this instance, emphasizing the earlier point, 'using a blockchain means that malicious third parties cannot access or interfere with any part of the system without disrupting the system as a whole and thence triggering a security lockdown' (Bradley, 2019). With the networking of the home space, home security becomes a question of data security. Here we can see how the aim for a more knowing and predictive home brings with it a focus upon how the data that inform that knowing are preserved and insulated. It also brings certain fears to the surface.

Echoing such fears and data anxieties, elsewhere these security issues have been described as perceived 'pain points' within networks of devices:

> As IoT is an emerging field today, still security issues are its pain points. When it comes to blockchain, it paves more security with its cutting edge technologies. So integrating this Blockchain in IoT … unambiguously creates an excellent framework that can never be hacked. … Blockchain with IoT render answers for the challenges like

a single point of failure, scalability, privacy, time stamping, trust, etc. (Osiz Technologies, 2020)

The need for a stable security barrier is the image this begins to conjure. Blockchain is presented as a solution, or a cure perhaps, to those 'pain points'. This then is a question of how the data accumulating within these connected spaces can be shielded and how the potential for such a 'single point of failure' is to be avoided. It is the interaction of components in these connections, the points of contact and interfaces, that represent these possible points of failure. From this perspective the internet of things almost becomes a series of potential weak spots from which data might be extracted or removed. Within this wider set of fears of a data-intensive and, therefore, hackable home, blockchain is projected as the means to secure or alleviate the 'pain points'.

As with the art market, the distributed and diffuse features of blockchain are again used to lend it legitimacy. As was observed in the art market, this is security through the diffusion or decentralization of storage. With the home, this is presented in terms of the decentralized ledger that requires no outside involvement in the record keeping. It is suggested, for instance, that 'third-party involvement is not required and the ledgers can't be altered without the network approval' (Osiz Technologies, 2020). The notion of the third party here resonates with the earlier discussion of the removal of 'gatekeepers' and 'middlemen' in the art market. Again, the sense of security comes from the home space not being placed into the hands of human actors but instead being applied by an automated and decentralized system. A sense of security is achieved by bypassing human involvement in connectivity and data storage. The aim to 'fully automate' the home returns us to the idea of the removal of human intervention and the exclusion of the need for any form of intermediation. Indeed, replicating the terminology used when blockchain was applied in the art market, we even see direct references to these so-called 'middlemen'. In this case it is claimed that 'blockchain technology and smart contracts does not need the middleman to verify transactions, enforce contracts and perform background checks' (Osiz Technologies, 2020). It should be noted that these types of algorithmic 'smart contracts' are not restricted solely to the art market, but are now being implemented on top of blockchains – such as non-fungible tokens and decentralized finance – to 'enable autonomous transactions to be performed without human intervention' (Chapman, 2021: 41). What is important here is the idea that an automated and distributed system is seen to represent a more secure future in which the human actor's vulnerabilities and inefficiencies are bypassed. As with the art market, the algorithmic thinking embodied in the notion of the smart home is based upon a notion of the unalterable and decentralized archive in which the perceived vulnerabilities of human involvement in data processes are managed through automated storage processes. As such, as with the art world, this discussion of the blockchain-managed home space draws upon or imagines a form of what might be thought of as *posthuman security*.

Hashed trust

In their analysis of the 'social construction' of blockchain across both mainstream and specialized media, one of Chow-White, Lusoli, Phan and Green's (2020: 12–13) key findings was that blockchain was constructed as a 'revolutionary technology'. In short, blockchain has come to be expected to bring about disruption, change and revolution. It is framed in these very terms and, with the implementation of the technology, it is expected that it will inevitably bring change with it. Blockchain is directly imbued with a sense of the transformations it will bring (or with the promises of an algorithmic new life as discussed in Chapter 1). This vision of change, Sandra Faustino (2019: 486) has found, can be located in the types of metaphors that accompany blockchain, some of which have arisen in this chapter. With particular ideas of revolution built into its very framing, the question then becomes what these expected revolutions will be, the logic underpinning them and the change that might then come about.

In his history of data processes, Colin Koopman (2019: 4) points out that where we have become 'informational persons', in which we are 'fastened' in place by these systems and categories, there is a deep fear of our data being deleted. Deletion would represent an unfastening. In a data-ordered society the deletion of data conjures significant fears, Koopman has observed. The reason for this is that once we have become our data, in Koopman's phrasing, then a lack of data will inevitably make it hard to function or operate. Indeed, in the context of social ordering based around data, it is perhaps inevitable that there will be acutely felt concerns over the integrity and accuracy of *the records.* As such, the rise of data processes – along with their intimacy and sheer ordering powers – has created the potential for new problems and new insecurities. Trust and preservation become the focus as organizations and industries pursue new 'immune mechanisms' that facilitate a 'new mode of immunity' (Anderson and Stenner, 2020: 99). The risks of data processes draw an immune-type response. The accounts of blockchain in both the art market and the smart home are illustrative of these types of immune mechanisms.

In a key article on blockchain and trust, Angela Woodall and Sharon Ringel describe the kind of 'archival discourse' around blockchain and the strong links it develops with notions of preservation. In line with Koopman's findings, they also point to the heightened concerns over data storage and archiving that come about when data are so essential to the running of the social world. Woodall and Ringel argue that there is an 'archival imaginary' of blockchain that is crucial in understanding its presence. Woodall and Ringel (2020: 2201) observe that this imaginary has expanded the reach of blockchain beyond the more obvious sectors, noting that technology providers 'have actively expanded their outreach to journalism outlets, academic institutions, government agencies, and archival organizations by marketing blockchain as a tool capable of preservation that is not only secure but also meets archival standards of authenticity, reliability, and trust'. The art market and the smart home are further instances in this bigger picture in which blockchain is pursued as a means for managing and securing the data archive. The tempting proposition in this 'archival imaginary' of blockchain is that it will provide a more trustworthy version of the record keeping that has hitherto been in place. Nigel Dodd (2018: 47) has noted that in the case of money, blockchain can reconfigure what trust is and where it is placed, especially where intermediaries are bypassed. Taking this into account, the case of the art market and the smart home fit into this picture of a change in trust and a repositioning of where that trust is placed – both case studies are also suggestive of this bypassing of intermediaries that Dodd observes. Further to this though, focusing on these two sectors also allows us to explore the particular embedded fears that exist and how a type of trust is built in blockchain, with its technical structures expected to overcome those perceived vulnerabilities.

Woodall and Ringel identify how certain characteristics of archiving become a central part of accounts of blockchain. These archival properties have facilitated its development as a tool across different sectors. It is by connecting blockchain with notions of the archive that trust is built. 'Archival imaginaries', they explain, 'capture a vision of what archives and blockchain should be and should mean, which pivots on imagined needs and technological capacities' (Woodall and Ringel, 2020: 2213). Evoking notions of the archive lends meaning to blockchain. These archival imaginaries, they argue, contain within them ideas of what should be and what those archives – in this case blockchain – might achieve. Crucial here, is how, as they identify, particular archival properties are evoked to generate trust in the technology and in its handling of data. They explain that 'establishing trust, in both archives and blockchain technology, the blockchain archival discourse creates what we call "archival imaginaries" – the desirable characteristics of an archive, not as it is but as it responds to perceived social needs in keeping with the technology's capacities' (Woodall and Ringel, 2020: 2208). The establishment of trust, they argue, is based upon the type of discourse and the form that the 'archival imaginary' takes.

Importantly, it is the way blockchain, through its archival imaginary, is perceived to respond to the requirements of society and to social needs that Woodall and Ringel identify as being of particular importance. In other words, it is not just the way the archive is understood or described, it is suggested that it is how those archival properties allow blockchain to be perceived to respond to prescribed social needs that is important in how trust is established and how blockchain is then pursued. This is something we saw in the fears that were perpetuated in the art market and the smart home and how blockchain was presented as a means to address those fears. In the case of the art market these were existing fears of fraud to which blockchain could be disposed, whereas in the case of the smart home it was blockchain's ability to secure the new connectivity of objects that was important. And so, along with notions of efficiency, there were specific prescribed social needs to which blockchain was then envisioned as a means of resolution.

Perhaps the most prominent prescribed social need concerns data security. Woodall and Ringel's (2020: 2213) point is that the ephemerality of data is important in seeing where the archival discourses around blockchain are active. Blockchain creates a stronger sense of permanency and reliability through its archival imaginary. In the current context, there is an imperative and a perceived need to locate ways to ensure the future availability and reliability of data. Woodall and Ringel (2020: 2213) found that blockchain was often 'vaguely construed' in favour of a focus on preservation. This means that it is not so much what blockchains do that matters, it is more a case of foregrounding what archival properties they bring. Woodall and Ringel identify notions of decentralization and distribution as being vital to the archival forms that are being imagined. In line with this, I have similarly found these properties are commonly foregrounded as the key properties of blockchain. The question implicitly raised by Woodall and Ringel's crucial observations concerns what is driving and fuelling blockchain's 'archival imaginary'. Once we know what the properties are and once we have acknowledged that blockchain is adapted to suit social needs, then we might wonder what logic is behind this and how the need for establishing trust has developed. We might think of that trust as being found within the decentralized blocks and hashes: a hashed trust.

Posthuman security and its draw

Interrogating the logics, modes of reasoning and ideals within accounts of algorithmic thinking can be potentially revealing. I have done this in the case of blockchain in this chapter and will continue to do this on different fronts in the following chapters. The developments covered in the earlier sections on the art market and the smart home could be thought of as depicting a kind of retreat or perhaps a leap into the apparent safety of the posthuman. Faced with expanding data processes and an attendant need to ensure that the records are not fallible, the posthuman becomes a potential space in which data anxieties might be resolved. Blockchain is both an embodiment of those anxieties and is sold as an antidote to the perceived vulnerabilities. And so various sectors, including the two described in this chapter, lurch towards a posthuman version of an algorithmic new life in which ordering arrangements are placed beyond the reach of human actors and into distributed chains. As this might suggest, with blockchain there is a kind of posthuman security at play in which the emphasis is upon the automation of records across distributed systems. A data-led society brings with it embedded fears of data security; there are inevitably going to be responses, especially as a future is envisioned in which more data then requires more security in order for the social world to function. The myths of the algorithmic new life are reinforced by notions of advanced data securities. The underpinning fear here is one of being misinformed by the data or of basing decisions and outcomes on data that have been altered. A loss of control was one way that this became embodied in the context of the data-rich and highly connected home space. And then we might add into this the way that posthuman security will be part of a response to the new risks that automation will bring and the ways in which automation is deployed to limit or control risks.

Even if we look solely at blockchain the vision created of greater humanlessness is not straightforward and we may want to reflect on how it might compare with critical and conceptual visions of posthumanism. Crucially, Rosi Braidotti has warned that 'the posthuman condition cannot be reduced simply to an acute case of technological mediation' (Braidotti, 2019: 3). Perhaps it is this very impulse that is a property of the dominant visions of blockchain in the two sectors explored in this chapter. Braidotti's observation is important because it captures exactly what is happening around blockchain and the reductive vision of the near future that it has a tendency towards. Part of what the concept of posthuman security might seek to do is to provoke an exploration of the type of posthuman visions that are embedded in the notions of a secure society, what they miss and how they narrow down the relations through which these technologies operate. The analysis of posthuman security should be about the impulses and means by which a version of the posthuman is being conjured. When blockchain is imagined, it is frequently pictured, to return to Braidotti's phrase, as an 'acute case of technological mediation'. A kind of humanlessness is taken to a point in which the human cannot intervene once the traces are in place.

We have a contrast here between the version of posthumanism embedded in understandings of technologies and those required for their understanding and critical analysis. Despite a stated awareness of the hype in some in these sectors, the powerful marketing and futurist accounts of blockchain, in which a secure society is imagined, potentially get caught up in this more reductive posthumanism to which Braidotti refers. Whereas, what an understanding of blockchain actually requires is exactly the more fine-tuned type of situated perspective on posthumanism that Braidotti is encouraging. Indeed, Braidotti has explained that the posthuman can be defined as 'both a historical marker of our condition and a theoretical figuration' (Braidotti, 2019: 1). The theoretical figuration of blockchain is already being exercised in how its coming deployments can realize potentials, outcomes and allay fears. This may be what we are seeing with blockchain; it is a posthuman future that incorporates both markers and theories, both implementations and visions. It is the competing levels of prominence of the human within the technological milieu that is particularly important and is where tensions may arise in these implementations and visions (which we will return to in Chapter 3).

Braidotti's approach in which the concept of posthumanism is both, as it was put, a condition and a theoretical figuration, creates a question over the role of that concept in any theories of posthuman security. Working with a similar separation, Pramod Nayar indicates that there are 'two important frames for the term' (Nayar, 2014: 3). Nayar explains the first of these two frames by suggesting that ' "Posthumanism" on the one hand merely refers to an *ontological condition* in which many humans now, and increasingly will, live with chemically, surgically, technologically modified bodies and/or in close conjunction (networked) with machines and other organic forms' (Nayar, 2014: 3). This indicates that the first frame is concerned with the lived experience of being posthuman. This is where the body itself can network or adapt to the environments and to changing technologies. This, Nayar points out, is how many people already live, to varying degrees, and how coming ways of life will be experienced also. It is important to note that Nayar is indicating that this is a combination of the current moment and of what will come.

The second frame outlined by Nayar is more focused on the active role of the concept of the posthuman and its analytical purpose. Nayar describes this second framing in the following terms:

'Posthumanism', on the other hand, and especially in its *critical* avatar, is also a new *conceptualization* of the human. Posthumanism studies cultural representations, power relations and discourses that have historically situated the human *above* other life forms, and in control of them. As a philosophical, political and cultural approach it addresses the question of the human in the age of technological modification, hybridized life forms, new discoveries of the sociality (and 'humanity') of animals and a new understanding of 'life' itself. In a radical reworking of humanism, critical posthumanism seeks to move beyond the traditional humanist ways of thinking about the autonomous, self-willed individual agent in order to treat the human itself as an assemblage, co-evolving with other forms of life, enmeshed with the environment and technology. (Nayar, 2014: 3–4)

In short then, posthumanism is both a lived experience and an analytical approach (the conceptual dimensions of this and the meshing of agency will be discussed in greater detail in Chapter 4). It is this second framing that indicates the type of insights that the concept might be geared towards producing. In particular, within this framing it is perhaps the closing line about enmeshing that is of particular relevance. It is to the question of the role of the human in technologies such as blockchain and many other advancing forms of automation that this type of framing guides us towards. It also might cause us to reflect on how an enmeshing with technology occurs and what forms of power are behind it, as well as the transforming hierarchies of agency that might afford it.

As such, for Braidotti (2019: 134) posthumanism captures or accounts for the specific intersections of agencies and argues that 'the human subject is therefore only one of many forces that compose the distributed agency of an event'. Posthumanism draws the analytical eye to the points of contact, the forces involved and the way that agencies are distributed. Indeed, the crucial phrase here is 'distributed agency'. With blockchain and many other automated and decentralized technologies, it is the way that the human is placed into such distributed agencies that creates tensions between the forces that compose their arrangements. One way to look at blockchain is in terms of the composition of its distributed agencies. In very blunt terms, the posthuman security attached to blockchain, driven by a set of fears, seemingly seeks to bypass or remove the human within these emerging relations of governance and ordering. In this sense, with the lack of trust creating a force that limits the human within these distributed agencies, blockchain could even be considered to be a form of 'radical posthumanism' (Gane, 2005). In the more specific case of the Bitcoin application of blockchain, Lana Swartz (2018: 632) elaborates the concept of 'infrastructural mutualism' to explore the type of 'cooperativist vision' of technology and society associated with these developments. It is the very decentralization in this 'infrastructural mutualism', Swartz (2018: 633) argues, that is appealing and that acts as the draw. This is suggestive of the pull towards an enmeshing of agency and towards the type of decentralization and distribution of these agencies that has been captured in the case studies covered in this chapter. Overall, we might conclude, it is the safekeeping of data outside human intervention that becomes the source of legitimacy, authenticity and trust within posthuman securities such as blockchain. What we see then is a posthuman security that may, we could consider, be fuelled by human insecurity. Blockchain embodies this sense of human vulnerability and the possibility for secure data.

Following this, the question then is about how the human is placed (or downgraded) into such a set of posthuman securities. Braidotti is careful not to abandon the idea of the human in the account of posthumanism that is being proposed. This contrasts sharply with what we might think of as the types of *commercial or technological posthumanism* that are attached to blockchain. The absence of the human is a central and defining part of that commercial posthumanism. Rather, Braidotti proposes that the analyst should aim 'to reinscribe posthuman bodies into radical rationality, including webs of power relations at the social, psychic, ecological and micro-biological or cellular levels' (Braidotti, 2013: 102). It may be the case that the image of humanlessness in blockchain needs to be analysed by bringing the human and human bodies back in – with blockchain the human is, of course, never really entirely absent. Braidotti's suggested approach to posthumanism is focused instead upon grasping the multi-scale and highly enmeshed relations of the social world and may require a challenge to the images of distributed agency that are perpetuated. Examining blockchain requires going against the grain of the sector's accounts of the technology, which seem to minimize if not eradicate the human from these systems, so as to uncover the tensions, webs of relations and forces encountered. It requires the human to be reinscribed where it might appear to be absented. The issue with blockchain is the particular way in which the abstention occurs. The type of security being pursued gets its authority from putting the human under erasure. It is the way in which absenting the system of human presence produces certain outcomes that are seen to be desirable and, crucially, more secure, that itself needs analytical attention. Indeed, when it comes to this dialling-down of the human, the establishment of an apparent humanlessness or the erosion of human agency it is the power of a sense of vulnerability combined with the security brought about algorithmic bypassing that is vital. It requires powerful visions of both security and insecurity in order for a move towards humanlessness to be presented as a viable future (we will reflect further on the opposing forces, tensions and responses to this in Chapter 3). Perhaps it is the very desire to limit the human that provides a focal point for analysing blockchain's posthuman securities (that very desire is a part of the will to automate that I will discuss in Chapter 6).

The priorities of deprioritization

As I have explored here, the posthuman securities incorporated into blockchain's futures are geared more towards the deprioritization of the human within these particular 'data assemblages' (Lupton, 2020). Perhaps what we have seen with blockchain is a kind of posthuman vision that, as was a tendency that Katherine Hayles (1999: 5) once observed, 'erases' the body and that tends towards 'an emphasis on cognition rather than embodiment'. Another way to think of this would be to follow Louise

Amoore's (2019) suggestion and explore a 'posthuman mode of doubt' by bringing 'doubtfulness' into the examination of algorithms and posthuman ethics (see also Amoore, 2020: 150–3). This would be particularly potent given that the posthuman security of blockchain is based on the perceived eradication of doubt. It would be revealing to examine the presence of doubt in algorithmic thinking, especially where posthuman securities, such as those examined in the case of blockchain, are aiming at the very removal of doubt. The very doubtlessness attached to blockchain then itself may be the focus for exploring its presence and future, and also may drive towards the non-erasure of embodiment and the reincorporation of aspects of the data assemblage that may otherwise be overlooked.

I've focused upon blockchain and the algorithmic management of data security in this chapter; there are, I'd like to suggest, a range of different types of posthuman security. This chapter has used blockchain as a focal point to tease out one version of this, but this should be understood in a wider context in which security is a major preoccupation of algorithmic thinking. These posthuman securities can occur anywhere but they inevitably tend to be located at the points where security is considered to be paramount or is considered to be a priority. At these points there will be forces intervening in the enmeshing of distributed agencies. In the case of blockchain this was about data security, but there are other instances where different automating processes are charged with posthuman securities. The list of these will be very long and can be located at the moments in which roles and responsibilities are passed to algorithms. Such an extensive list would include things that are invisibly integrated or that occur in quite mundane forms, through to more obvious and visible forms, such as automated and self-driving vehicles including ships, cars and other logistics and robotics. It might also be found though in fields such as financial trading and risk management (see Chapter 3), as well as being tied into other types of data management and storage. Algorithmic thinking is always likely to have a security dimension. In these many and very varied instances the human can be seen as being associated with insecurities to which automation is presented and pursued as a resolution – this becomes about the management and control of human input within the management of potential fears and risks. In other cases posthuman securities might be about securing wealth or value. The speed with which automated systems are able to trade, for example, gives them the edge and therefore aligns them with a different type of posthuman security, one in which value is secured. There are traces here of three versions of posthuman security – in data, mobilities and wealth – that could be further explored and that might be added to. In other words, by focusing in upon blockchain and data security I have only begun to elaborate on posthuman securities as a broader concept that might allow for an examination of the forces of humanlessness in general. As we will see in the following chapter, this force is in tension with a push towards the integration or reintegration of the human. Posthuman security is itself a site of tension.

Overall, as an illustrative case blockchain is suggestive of how a kind of posthuman security thrives upon human insecurity. Katherine Hayles has pointed out that 'we need frameworks that explore the ways in which technologies interact with and transform the very terms in which ethical and moral decisions are formulated' (Hayles, 2017: 37–8). Such a research agenda is needed to examine blockchain as it continues to be rolled out. This agenda should seek to capture and examine the promises and potentials as they are envisioned as well as the various implementations, applications and adaptations that occur, alongside the shifts these bring to organizational practices, hierarchies, values and relations. This approach might seek to examine the underpinning sources and points of insecurity and vulnerability as well as the projected means by which technologies like blockchain present themselves as adaptable to solving those insecurities. This is about the fears as well as the possibilities, the desires as well as the implementations. It is also about the reasoning that occurs in the processes of disembodiment that automation and algorithmic systems might be aimed at bringing about.

Blockchain's presence and the vast hopes placed upon it tell us something about how people feel about data and hints at the senses of vulnerability that are experienced in a data-ordered society. What I have suggested here is that blockchain represents a form of posthuman security that seeks to address a sense of human insecurity. Understanding the tensions this creates will be fundamental to understanding the ongoing implementation of blockchain and related technologies. The use of posthuman securities does not end with blockchain. Rather, this is one technology and one form in which it is present; it will no doubt be of greater significance and potentially produce stronger tensions in more advanced forms of automation. Posthuman securities are likely to be found anywhere that algorithmic thinking is promoted as being the answer to our vulnerabilities. The question posthuman security poses concerns what exactly is being secured, how it is being secured and why the human and algorithmic relations are being depicted in the way that they are. It is not then a concept for analysing only the notion of security itself, it is also about the underpinning impulses towards particular types of posthumanism. Posthuman security calls for a very specific type of examination of humanlessness in algorithmic thinking, one in which particular fears and perceived vulnerabilities become a part of the advancement and development of visions of a secure society. As we look ahead into the myths of a new life defined by advancing automation the question of posthuman security is likely to rear up in front of us, especially as we think about the many forms of automation and algorithmic thinking that are based upon bypassing the human or, at the very least, are focused on reducing human agency within the mix.

3

Overstepping and the Navigation of the Perceived Limits of Algorithmic Thinking

Published in the 7 March 2020 edition of the *Financial Times*, the private banking firm Investec ran a full-page advert that seemingly sought to defend human decision-making. Going against the apparently relentless tide of automation, it gestures towards the pitfalls of an irrational and inflexible form of algorithmic thinking. The advert poses what appears to be a rhetorical question: 'Who would be most likely to grant you a mortgage? An algorithm? Or a human being?' In the unlikely case that the reader is unsure as to their position on this question, the background is filled by a monochrome photo of a comfortably seated human – it is not clear if they are the imagined customer or a representative of the lenders.

This particular advert is suggestive of two related things. First, by responding to it directly the advert highlights the established materiality of algorithmic social ordering. In order to function its key message requires there to have been some form of existing encounter with algorithmic structures of some sort. Second, it is indicative of the way in which the very notion of the algorithm has moved into public consciousness (as discussed in Beer, 2017). The apparent rush towards *being algorithmic*, in which organizations seek to present themselves as devolving powers to the apparent neutrality, objectivity and heightened efficiency of algorithms, creates opportunities for others to present themselves as providing an alternative. This is not an alternative to *being algorithmic*, I would add, it is more often simply a different version of it. In other words, the push towards algorithmic properties creates a space in which the human can be knowingly and actively reinserted into these systems. In the case of the Investec promotion, the attempt is to appear to *be algorithmic* while not abandoning a sense of human values. In other words, it is an attempt to actively seek to present this as an organization that uses automation without appearing to be *too automated*. There is an active avoidance of that particular boundary. Investec are, it would seem, aiming to avoid overstepping the perceived limits of algorithmic thinking. That is to say that there is an attempt to remain within the boundaries of what might be considered to represent the acceptable presence of algorithms. These limits are based on what is understood to be the right type of mixture of agency and the right type of positioning of algorithms within the division of labour.

In its depiction of the choice between human- and algorithm-based decisions, the Investec advert calls upon an already existing tension. It is constructed around an assumption that people are worried that algorithms are inflexible and unaccountable. The advert relies upon an existing feeling that algorithms have too much control; it is playing to this type of concern that human discretion is being swallowed by automation. This positioning continues into the detail of the advert. In much smaller text and in more conclusive terms, the Investec promotion adds:

> When you need a mortgage, ask a human. For while we find algorithms very useful, they are not known for their flexibility. For those who define success as living and working on their own terms, help from a kindred spirit counts for a great deal. So, if you prefer talking to humans about a mortgage, try one of ours.

In terms of its positioning, this represents a fairly gentle defence of the human and of human discretion. It is a defence based on a notion of human flexibility. It is far from a complete rejection of algorithmic thinking nor is it an erasure of algorithms. It is instead a statement of a certain form of algorithmic thinking, one in which the human is brought closer to the surface and one in which the algorithm is retained where it is deemed useful. A kind of algorithmic pragmatism perhaps. The utility of algorithms is noted, while the sharp binary cuts and unbending decisions of algorithmic thinking are challenged in favour of something warmer, more responsive and that offers a greater sense of empathy. The inflexibility of algorithmic binaries are being contrasted with the apparent flexibility of human thinking. Bringing the human back is a way, it would seem, of seemingly easing the tensions.

This particular advert, it could be concluded, is aimed at those potential clients who would prefer to feel that algorithms do not exert complete or unyielding power over their lives or over their opportunities, chances or outcomes. There is a sense here that there are those seeing this advert who do not wish to be known or judged solely through the algorithmic analysis of their data but who, instead, want some sort of rapport or human interaction to play a part in those decisions. At least that is the framing of these systems; they are framed in terms of a set of concerns. The aim is clearly to appeal to those who would rather see the algorithm categorically in service of the human decision-maker rather than the reverse (as I will go on to discuss later in this chapter). The advert aligns Investec with that imagined audience. The message is that this provider makes mortgage decisions based upon experience and know-how, rather than leaving it solely to the cold calculations of machines. Here it is the inflexibility of algorithmic decisions and their cut-and-dried certainty that is presented as being the problem. Such fixities are juxtaposed with something more open and persuadable: the human. This advert positions Investec as leaning against the type of humanlessness discussed in Chapter 2.

Yet, as we see, even here algorithms are not abandoned entirely; they remain, we are told, 'very useful'. This is a kind of functionalist-type algorithmic thinking that is built around notions of usefulness and utility. This isn't the complete removal of the algorithm as such; instead we just have a different attempt at *being algorithmic,* one in which the human and algorithm are entangled, with the balance placed towards the human in these impressions. It is neither an attempted erasure of the human nor of the algorithm – instead, it highlights and exists within the tensions between them. With its reliance on an existing understanding of what algorithms might be and the type of decision-making processes that they might be a part of, it could be concluded that this promotional advertisement requires a well-established version of what Taina Bucher (2017) has called the 'algorithmic imaginary'. Bucher uses this term to reflect on how an understanding of algorithms becomes a part of the way algorithmic systems are understood and approached. Indeed, within this imaginary it is not uncommon for developments in automation and AI to be accompanied by deep-rooted concerns about discretion (see Hall, 2017). As I will explore in this chapter, the algorithmic imaginary described by Bucher has edges; there are limits of what is possible or what are considered to be the acceptable extents of automation contained within it. There are also ideas about what those limits might be, how they might be crossed and what might be beyond that particular iteration of algorithmic thinking itself. The perceived limits are, of course, contested and in tension. With this is mind, and having seen this snapshot of algorithmic positioning in process, we might then ask what the perceived limits of automation are and how they are being defined and redrawn.

The various dimensions and forms of algorithmic thinking mean that these limits are far from straightforward or easy to track. There is no simple and settled consensus to look at. The limits of algorithmic thinking are being constantly contested and redefined. The range of ways in which the limits of algorithmic thinking are being redrawn makes their analysis something of a challenge. As well as being in flux, these lines are relational, dynamic and open to contestation. I would go as far as to say that the perceived limits of automation – the multiple lines and boundaries that are drawn around its acceptability – cannot be fully known. Yet, at the same time, it is possible to create snapshots of these processes. Various approaches and focal points, such as the Investec advert, may be utilized to provide glimpses into these limits that might then reveal something of the wider processes and forces of which they are a part. What I focus on in particular to try to make this visible are the active and often implicit attempts that are made to navigate the limits of algorithmic thinking. In particular, I focus here on the role that a notion of *overstepping* plays in the maintenance as well as the breaching of these perceived limits of automation. Such a focus provides insights into the drawing of limits and into how a notion of *too much automation* can be wrapped into the very expansion of algorithmic thinking. This chapter is about the sensitivities expressed towards the limits of algorithmic thinking and how those sensitivities are actively deployed and managed as a part of its expansion. A knowledge of the limits becomes in itself a means for their navigation.

Using a case study of the widely used and highly influential BlackRock asset-management system Aladdin, this chapter explores the implicit circumscription of agency that occurs as organizations seek to navigate the perceived limits of algorithmic thinking. Using this particular example as a suggestive instance of how we might grasp these limits, the chapter looks at how organizations subtly seek to *become algorithmic* without seeming to entirely hand over the decision-making to algorithms. The chapter looks at the framing of these systems in the commercial presentation of a particular system, approaching these framings as a part of a 'dispositif' (Foucault, 1980: 194), so as to make visible an instance in which the limits of automation are being worked and reworked. Alex Hall has argued that 'with the diffusion of technologies into everyday decisions about how we are to be governed – from policing and finance, border security and health – the outline of the discretionary decisionmaker is in flux, even if the problems of discretion ... are perennial' (Hall, 2017: 502). Taking Hall's argument, we might then explore how long-held issues around the understanding of discretion take particular forms within these expanding algorithmic systems and their attendant modes of thinking. Drawing out the specific framing of a particularly powerful system, this chapter looks at this flux and at how discretion and the roles of human subjects are actively being remade in this context. As such, this chapter is concerned with how the very modes of reasoning behind decision-making feed into and enable the expansion and development of automation.

Within the broader tensions of algorithmic thinking, this chapter looks at how human and algorithmic relations are implicitly and quite subtly projected onto these processes and at how legitimacy emerges from these relations. It reflects on how this legitimacy is maintained through attempts to avoid *overstepping* established or perceived limits, illustrating how a sense of the boundaries can inform and be a part of algorithmic thinking itself. The case study used here is illustrative of how the human is retained when automated systems are envisioned. It explores in particular how the human is placed into particular roles in this circumscription of agency and how the category of the human is then placed within what is imagined to be a unified and collective system. The chapter argues that the depictions of the limits of algorithmic thinking, which are presented in a kind of balance with aspects of human intervention, are an active part of the deployment and expansion of automated processes. I will suggest that the development and further integration of automation is based upon an understanding and vision of what is felt to be *too much automation*. As such, the concept of overstepping might be used to think about where these limits are drawn and how they are being actively managed, adjusted and pushed back. The avoidance of overstepping is part of how automation spreads and expands. As such, this concept of *overstepping* may be used to examine the active processes involved in navigating the perceived limits of algorithmic thinking.

Establishing the algorithmic

A product of the financial technology and service provider BlackRock, Aladdin can be understood to be an influential risk-management software. Although this label could be considered to be a little too reductive when considering its features and functions. Operating in 'real time', Aladdin has been described as a 'platform that helps in decision making, effective risk management and efficient trading' (Mauricio, 2017). Instantly, we can see here an echo of ideals of speed and 'real time' that populates discourse around data-led processes (see Beer, 2019a). Back in 2014 it was suggested that Aladdin was already 'more powerful in some respects than traditional politics' (Curtis, 2014). That would, of course, depend on what is meant by power, yet this type of observation at least begins to build a picture of the type of influential calculative presence that Aladdin is imagined to possess within financial structures. Even if these types of comments overstate the case, they are still suggestive of the sway that Aladdin is thought to have. In a 2014 blog for the BBC website, the filmmaker Adam Curtis (2014) described how Aladdin had:

> within its memory a vast history of the past 50 years – not just financial – but all kinds of events. What it does is constantly take things that happen in the present day and compares them to events in the past. Out of the millions and millions of correlations – Aladdin then spots possible disasters – possible futures – and moves the investments to avoid that future happening.

There is a version here, in this use of future horizons to shape the present, of Louise Amoore's (2013) accounts of the 'politics of possibility'. In this description, past events and the data gathered about them shape the present and are also impacting on foresight. Back in 2014 Curtis was concerned both with the amount of involvement that Aladdin already had in decision-making and also with what the type of power that this system held. As well as giving a sense of scale, Curtis' comments would also indicate that the context in which Aladdin developed included vocal concerns over the power of the automation it facilitates, especially given the reach it holds. Such concerns are likely to become a part of how such technologies are framed and presented, with responses to outrage and fears becoming part of the framing of such technologies as they continue to develop. And we might begin to see here how the responses to technologies may feed into an understanding of the perceived limits of automation.

The Aladdin acronym stands for asset, liability, debt and derivative investment network and it 'began as a simple ledger for bond portfolios shortly after BlackRock was founded in 1988' (Henderson and Walker, 2020: 11). This represents a fairly long history of development for software of this type. The wider context of the expansion of data and data mining is important within this too. Aladdin found a foothold as companies sought to use their data and as the art of decision-making was reimagined in terms of the seemingly heightened rational judgements of automation (the wider context of this is described in Andrejevic, 2020: 62–8). After a series of software extensions for a limited number of its clients, BlackRock released Aladdin as a standalone product in the year 2000, some 12 years after its initial development (Henderson and Walker, 2020: 11). Aladdin slowly built up its availability and functionality, its user base and usage then increased relatively rapidly after the point of wider release.

Since then Aladdin has grown to become a very significant presence in the implementation of data processes. Back in 2018 the 'platform' was reported to be supporting portfolios of 'more than $18 trillion' and had been 'adopted in some form by 210 institutional clients globally' (Whyte, 2018). In terms of its scale of use and influence, it has been claimed that:

> The total value of assets under management by BlackRock is $6.3 trillion. But Aladdin also delivers risk analysis on the assets managed by its clients, which are valued at more than double that amount. Overall, Aladdin has an effect on the management of around ten per cent of the world's financial assets, or around $20 trillion. Over 25 years, it has grown into a system that is directly or indirectly responsible for more than four times the value of all the money in the world. (Dunn, 2018)

Aladdin is deeply enmeshed in global finance. That type of scope and reach is almost unimaginable, yet it is suggestive of just how significant Aladdin has become and is also indicative of why some of the claims about its apparent power might have weight. Aladdin is a major part of the management of finance and risk, and therefore is an important focal point for understanding the expansion of algorithmic judgement and automation. Aladdin is operating on a vast scale and has roots that are embedded in the underpinning structures of contemporary capitalism.

Aladdin's automated functions inevitably raise questions about discretion, decision-making and the combination of human and machine agency. The size of its potential influence means that such questions carry a particular poignance for software and systems of this type. The 'intimate entanglements' (Latimer and López Gómez, 2019) of agency inevitably become a part of the accounts of the services that Aladdin is depicted as providing. My suggestion is that it is because of the emphasis upon the particular and implicit entanglements of agency that Aladdin is an important example, not least because it is so widely used. Alongside this, it is also an important example because its success means that it will potentially offer a model by which others may also present their own algorithmic thinking. As with the Investec example with which I opened this chapter, BlackRock also appear to be actively aiming to avoid overstepping and could be understood to be actively negotiating the perceived limits of automation in the way that they frame Aladdin. Like Investec, Blackrock gently steer away from any idea that they may be providing *too much automation*. Instead they place the human into certain roles within these automated systems – a circumscription of agency is a part of the framing of Aladdin. In other words, there is an underpinning notion of overstepping that seems to inform the navigation of the limits of what is deemed appropriate in automation. Algorithmic thinking concerns itself with how its own limits might be managed. This sense of what overstepping would look like and its potential consequences is woven into the rationality and projection of automation processes and algorithmic systems.

The potential problems that might come from *too much automation* are a part of the discourse in the sector. For instance, highlighting a particular concern, a recent report has observed that:

> Though Aladdin does not tell asset managers what to buy or sell, some argue that if a large portion of global assets respond to the warnings that Aladdin gives off, trillions of dollars will react to events – such as the outbreak of a pandemic or war in the Middle East – in the same way, causing dangerous herding behaviour. The more investment managers and asset owners rely on Aladdin to gauge risk, the less responsible they become for their portfolio decisions. (Henderson and Walker, 2020: 11)

This captures a concern that too much automation might lead to algorithmic herding behaviours. This actually reflects the wider 'fears of herding' that

Christian Borch (2016) has identified as being a dominant trope around financial technology in his analysis of algorithmic finance and the 2010 flash crash. Henderson and Walker's report also highlights how concerns over the pitfalls of automation accompany the discussion of Aladdin, to which then its commercial framing is likely to respond. The concern expressed in the report is that an algorithmic rationality can lead to similar decisions being made in numerous places. The result of this is depicted as a kind of swarming or herding that might occur as the automation leads to a reactive mode of algorithmic thinking in which the crowd is simply followed. Instructional rather than dominant, the descriptions of Aladdin that I will go on to explore depict it in terms of the management of prescription and control in decision-making. As I have already argued, and as we begin to see in the case of such swarming, notions of overstepping are an integral part of algorithmic thinking itself.

To avoid the herd outcome mentioned previously, the commercial framing of Aladdin is based upon its ability to break with patterns and focuses instead on the different levels of responsibility for decision-making placed within the organizations in which it is operating. As with the Investec advert, the question that is implicitly asked within the framing of this software is what balance of human and machine is imagined to produce optimum decision-making (or to appear to do so). Here the type of tensions around the human and humanlessness run from Chapter 2 and into this chapter. The concerns over herding are themselves illustrative of the types of perceived limits of algorithmic thinking that require navigation by those who wish to expand the reach of such systems. It is implied by BlackRock that it is human intervention that would stop such algorithmic herding and that can break patterns established by more automated forms of algorithmic thinking. With this implication as subtext, as we will see, the framing of Aladdin is suggestive of how in these depictions the human is retained in a particular role – a circumscribed agency – in order to manage the problems of herding or the potential for automation to lead to the reactive following of patterns. Here we begin to see how agencies are actively meshed within such algorithmic thinking.

Given the scale of its use, Aladdin is a potentially powerful presence in the assessment of risk and therefore in decision-making, it is also then a useful case study for thinking about the presence of algorithmic thinking and for looking at the navigation of such perceived limits. In terms of its background, Aladdin has moved quickly to this position of relative power. As Henderson and Walker (2020: 11) have described:

> The system has expanded rapidly. Blackrock's 2015 deal to acquire FutureAdvisor spawned an offshoot of the platform for financial advisors today used by Morgan Stanley and UBS. BlackRock also offers

a version of Aladdin for custody banks, including BNY Mellon, that safeguard assets managed by fund groups like BlackRock. Last year, BlackRock acquired Efront, a private equity tech platform, for $1.3bn, extending Aladdin's reach into less liquid assets.

This is a story of relatively steady growth through acquisition and implementation. This particular software is closely aligned with transformations in the financial sector and a series of related organizational transformations. Here we can see how the establishment of automation is not produced in a vacuum but is an outcome of wider forces, ideals of automation, iterative relations between actors and even a set of organizational shifts and changes. In this case, Aladdin's form and functionality have mutated in response to acquisitions and the shifts of an emerging customer base. Notions of risk and of the potential of data and algorithms are also a part of this story. As such, algorithmic developments should be seen through such a lens of organizational change.

Aladdin's expansion is illustrative of some much broader changes around data and automation in particular. Henderson and Walker (2020: 11) suggest that some 'powerful trends have buffeted Aladdin's rise'. In particular, they add, 'investing has become more electronic and reliant on big data. As the tools that process the information have become more complex, investors, fund managers and insurers have turned to larger platforms such as Aladdin to replace multiple specialised systems.' The expanding scale of data has been understood to require larger-scale systems to manage those data (as was discussed in Chapter 2). As a convergence technology, Aladdin is understood to address a broader pursuit of automation in decision-making processes and in data management. Beyond this, we could also add that the expansion of systems like Aladdin is driven by the pursuit of an ideal of a perfect and perfectible decision-making process in which a balance of automation occurs and in which data are seen to offer new opportunities for accuracy and vision. This then is an illustration of the establishment of algorithmic thinking in action.

Life cycles and end-to-end logics: setting the spatial and temporal limits of algorithmic thinking

Turning to BlackRock's own accounts of Aladdin and their operationalization of a particular 'algorithmic imaginary' (Bucher, 2017), we find a mixture of technical functionality and rhetorical projection concerning its capabilities. These accounts contain subtle notions of how these systems are able to enhance decision-making. For instance, the sense of an ability to answer questions through the vast available data echoes in claims such as this one concerning how Aladdin allows the user to:

Rapidly test thousands of potential scenarios every day – answering questions like, 'How will inflation affect me?', 'What impact will a change in oil or gas prices have?', or 'What happens if there's a recession in Europe?' – that help you anticipate, interpret and respond as changes – big and small – happen across the world. (BlackRock, 2020a)

The use of future scenarios to shape the present, as discussed by Louise Amoore (2013), is a key part of the framing of this type of data usage and its automated analysis (and for a discussion of this in terms of data analytics see Beer, 2019a). This is expressed in terms of anticipation in this passage. This brief passage is also furnished with relatively common claims about the vast scale of analysis associated with the revolutionary potentials of 'big data' (discussed in Kitchin, 2014). This is embodied in phrases such as: 'Access hundreds of risk and exposures metrics' (BlackRock, 2020a). As well as inferring scale, there is an anticipatory logic to this type of algorithmic thinking. The ability to foresee is introduced in this account as a way of acting before or in advance of events (see Beer, 2019a: 27–9). There is then a predictive sensibility within this type of algorithmic thinking, a rationality based on what will happen next and how it might be brought into view through an analysis of the data.

Central in these accounts is that Aladdin is *more-than* it might be anticipated to be; it is presented as *more-than* a restricted piece of software and *more-than* lines of functional code. Bringing to the fore the ideas that algorithmic thinking can adapt and respond, rather than it being framed simply as a fixed tool, Aladdin is envisioned as a more active and adaptable presence:

More than portfolio management software, Aladdin is an end-to-end operating system for investment professionals to see their whole portfolio, understand risk exposure, and act with precision. Through Aladdin, BlackRock is committed to helping clients navigate volatility and market uncertainty. (BlackRock, 2020b)

Reactive and anticipatory in its form, the tone of the description is both expansive and expansionary (for more on the continuing expansion of Aladdin, see Whyte 2020; and for more on the recent increase in environmental data used by Aladdin, see Basar, 2020). This is presented as *more-than*: more than mere circumscribed 'portfolio management software'. Again, this is suggestive of potential reach. Algorithmic systems are based in a logic that sees them as *more-than themselves* and as being beyond restrictions and limits (which I will return to in Chapter 4). These are framed as open-ended technologies that are under development and, as I have argued before, are based upon the idea that they are constantly mutating and evolving (Beer, 2019a). Algorithmic thinking also leans towards an image of being all-encompassing and seeking ever greater coverage. It would seem that the first step towards the notion of perfect decision-making is to be completist about the processes through which the decision is made. In other words, this notion of decision-making is rendered perfectible through the inclusion of all aspects from, as the phrase goes, *end-to-end*. A notion of complete control is incorporated into this. No aspect of the system remains out of reach or out of control. Algorithmic thinking fills this space, from end-to-end.

As well as incorporation and complete control, efficiency inevitably weighs heavily in these product framings. This echoes the focus on efficiency of data storage discussed in Chapter 2, but in this case the focus is a little less algorithmic. In the following the completism and end-to-end logic mentioned earlier are embodied in a notion of a decision-making *life cycle*:

> Aladdin is investment technology that brings efficiency and connectivity to institutional investors and wealth managers. The same technology that BlackRock relies on for investing, Aladdin provides clients with a common language across the investment lifecycle in both public and private assets and enables a culture of risk transparency among users. Aladdin technology empowers investors around the world to run with clarity. (BlackRock, 2020b)

The notion of efficiency and the completeness of the data incorporated into this decision-making is the crucial thing to take from this. Where the notion of the end-to-end provides a spatial set of limits, here the life cycle offers a more temporal dimension. The life cycle of a decision is embodied within the assemblage. It is here that we find the core ideals around the automation of perfect decision-making being established and communicated; it is also an illustration of a type of algorithmic thinking in process. Although, of course, there are obviously limits to algorithmic decisions and to the type of transparency being spoken of here (see Flyverbom, 2019: 136–8), they may still be envisioned as being mobile and as stretching these spatial and temporal boundaries. This type of positioning around efficiency continues, with claims such as: 'With a single, standardized data set, Aladdin brings clarity, efficiency, scalability, and cost savings to the entire investment lifecycle. Aladdin's native risk analysis makes Aladdin a truly end-to-end platform' (BlackRock, 2020c). The life cycle appears again here, as do the core principles of efficiency and clarity in this apparently complete end-to-end logic that is said to be offered. Clearly, and perhaps unsurprisingly, algorithmic thinking centres on efficiency. What is perhaps more notable here though is the combination of the end-to-end and the life cycle upon which this particular type of idealized efficiency is based.

The algorithmic factory: visibility, attention and oversight

Inevitably, perhaps, two key things emerge as this type of end-to-end logic is conjured. The first is the presence of a type of analytical thinking. The second is the presence of the infrastructure that underpins that analysis. This is envisioned as an infrastructure of data hosting and access. In one overview of the technology these two streams are captured together:

> Delivered as a hosted service, including the Aladdin technical infrastructure, system administration and interfacing with data providers and industry utilities. BlackRock operates a data and analytics 'factory' with 600+ professionals focused on creating and quality controlling data and analyses for clients. (BlackRock, 2020c)

In the analytics factory described here human oversight is inserted back into the system that is being described. This is where the limits of algorithmic thinking are being drawn and where effort is being taken to build a picture of a system that isn't *too* automated. This algorithmic factory is populated with humans. There are traces here of the kind of overstepping that is embedded in algorithmic thinking. The impression of human intervention is built in so as to avoid the vision of automation beyond the human (and is thus in tension with the processes described in Chapter 2) and to ensure that the limits of acceptability are not overstepped. The type of analytics being outlined are based upon extensions of visibility and mobility. The assemblage is promoted to the foreground in this account of automation; it is a 'data assemblage' (Lupton, 2020) that retains aspects of the human. Indeed, it retains a large gathering of human actors within that assemblage. I'd suggest that these represent subtle traces of a circumscription of distributed agency. In a disjuncture with images of an embodied data assemblage (see Lupton, 2020), it is instead an arrangement in which the human is depicted as playing a partial and circumscribed role. This is encapsulated in the use of terms such as 'interfacing' and to a lesser extent 'infrastructure'. Within this, the limits of algorithmic thinking are sketched out to retain visions of human oversight and quality control. This is a type of oversight of the algorithm.

The other aspect of the building of the assemblage captured in the terminology is to be found in the analytic *factory* that is described – the use of the term factory in speech marks is instructive. In this factory large numbers of actors assist in the analytic process (something I will look at in more detail in a moment). It would seem that there is a retention of the human in the combined agencies being described. The legitimacy of the decision is forged in the idea that human actors are involved in quality control and in the oversight of the automation embedded in these systems. In other words, the pursuit of what is implied to be a perfect decision in this case is not one in which automation is envisioned as being unchecked. Rather this algorithmic factory is envisioned as a space in which the human is retained to oversee the actions of algorithms – the role is one of, as it is put, 'quality control'.

The analytics embedded within the infrastructures described appear to be based upon what this combination of human actors and technologies make visible. The pursuit of the perfect decision is also a pursuit of an infrastructure that is seen to afford heightened visibility. This is an automated visibility in which these systems are presented as prosthetics to human vision. Aladdin, it is claimed, is 'working … to help … to see risk more clearly'. The claim is that risk is made more visible. They add that 'Aladdin Wealth provides the transparency and clarity for you to understand and manage the risks in your business and client portfolios, enabling more informed investment decisions' (BlackRock, 2020d). The ideal of being ever-more informed is embedded in this, but the more pressing idea is the achievement of seeing risk more clearly. Algorithmic thinking is based upon the ideals of a type of heightened visibility. This clarity of vision, as we will see, is a repeated theme in the descriptions of Aladdin. Seeing into the depths is routinely presented in data analytics discourse as being a part of the analytical vision (as discussed in Beer, 2019a). Here the automated visibility of the depths is described in terms of a 'deep-dive':

> Portfolio Deep Dive: Analyze client portfolios the way they see things, from individual accounts to total household wealth, including external assets held away. Understand portfolio exposures across asset classes, geographies, risk factors, top contributors and sophisticated scenario analyses. (BlackRock, 2020d)

This multi-scalar perspective is presented as a crucial part of how the data are seen and then how they are analysed. Indeed, this volumetric notion of a type of sight that moves between layers is not uncommon in discussions of commercial data analytics (Beer, 2019a: 29). The deep dive into data repeats the frequent fluid metaphors that are often used to envision data and data processes (Lupton, 2015: 106–11) – in this case it is the focus upon a human-orientated algorithmic thinking that is said to facilitate these depths.

Alongside this idea of seeing into the depths, there is also a sense of responsiveness. With these accounts, seeing into the depths also facilitates decisions that are somehow responsive or even anticipatory. This is decision-making based on ideals of a heightened instantaneity or 'immediation' (Andrejevic, 2013: 146). You can, they say, 'leverage Aladdin's "one system, one database, one process" model and the comprehensive quality control services provided by the Aladdin team to help gain enhanced control and the reassurance that everyone is looking at the same high quality data in real time' (BlackRock, 2020a). We may note here again the mention of the 'team', which gently reiterates a form of human oversight within the assemblage. Furthermore, such ideals of quick and real-time responsiveness – facilitated by automated data analytics – are also reflected in the availability of 'alerts' that call for attention. A kind of data-analytic hailing is occurring in which human attention is intermittently drawn towards certain outputs:

> Alerts: Shift your business to be more data-driven with systematic alerts that identify clients and accounts requiring attention. Create rules to identify both issues and opportunities, allowing advisors to know which clients to contact and where to focus the conversation. (BlackRock, 2020d)

Part of this visibility is the way that the human eye is drawn by algorithmic processes. The attention is drawn, it is claimed, towards the information required for decisions to be rapidly made. This is also about a selective type of involvement. The notion of targeted information is the basis of this algorithmic thinking. In this account, human attention is called by the automated machine when required for discretion to be exercised. The machine finds what matters and informs the human; that is the model being suggested here. The interlocking of agencies of the algorithmic factory are pictured here in terms of a call and response.

Accuracy, control and oversight

In the case of Aladdin, the vision is of how risk can be comprehended and reacted to. Among all of the masses of data, in this framing of the software attention is channelled to selected aspects of risk. The temporality of the process is also at the forefront of the picture being created of its functionality, as is the perspective this timing gives on risk:

> With Aladdin, you have access to a thorough and timely understanding of risk. BlackRock's industry-leading risk models and multi-asset class analytics cover a broad universe of investment instruments – from fixed income and equities to real estate and hedge funds. Aladdin offers you not only a comprehensive view of your positions and risks but also helps you to zero in on exposures across every single portfolio and team. And because of Aladdin's global footprint and strong support across asset classes, Aladdin can support modeling for large and complex portfolios. (BlackRock, 2020a)

This is rendering visible exposure to risk. And so the promise is made of a 'comprehensive view', from which the analytics inform decisions over risk and risk management. The notion of scale and volume repeat in these terse summaries of the services provided. In this case we see reiterated the broadness of the algorithmically prostheticized view and the depth to which it can see. In terms of entangled agencies in this vision, the automated system is said to draw attention to where it is needed, selectively, while also facilitating a depth of vision.

In this type of algorithmic thinking human attention is envisioned as being a precious and limited commodity that the algorithm only periodically calls upon, when necessary. There is a sense here of the building of a legitimacy in the decision-making systems being deployed, with the algorithm turning to human actors for discretion where necessary. Such reassurances in automation and data are found in various places in the descriptions of Aladdin: 'Feel confident in the robustness of your analytics, as Aladdin monitors 2,000+ risk factors each day – from interest rates to currencies – and performs 5,000 portfolio stress tests and 180 million option-adjusted calculations each week' (BlackRock, 2020a). This builds a sense of volume and scale, which are often associated with ideals of accuracy and objectivity in big data (famously questioned by Boyd and Crawford, 2012). This reassuring voice here is saying that the software can build a sense of confidence in its users, once they are exposed to its scale of analysis. And so there is a subtle sense of overstepping and of limits being built into the discourses of algorithmic thinking.

Crucially, the perspective on data and risk begins to divide the automated aspects of decision-making, attaching it to particular parts of the process. In other words, the entanglement of human and machine agency takes on differing configurations at different stages in the process and is found in different combinations within the system. This gives a glimpse into the variegated nature of algorithmic thinking. In some places automation lends authority, as we saw in Chapter 2, whereas in others it is the greater and more active part seen to be played by human actors that is said to ensure the 'quality' of outcome. In the former, the algorithm takes the lead in ensuring authority by limiting human input; in the latter case it is the oversight of algorithms that is the human role. The positioning here is that the software *enables* whereas it is the human that *oversees*. For instance, the modelling and the creation of models are described as human-led with, as they put it, 'Aladdin's comprehensive suite of sophisticated models, built over the past 20 years by a dedicated financial modeling team' (BlackRock, 2020a). The models are not simply the domain of automated analytics; the human is reinserted in these practices of prediction.

Beyond these processes, as has already been highlighted, quality control is one such area in which the human appears to take on greater responsibility within these accounts, whereas in the initial stages the responsibility passes almost entirely to the automated technologies. I've already outlined how it is the automated analysis that draws human attention where required; this aspect resurfaces in these framings. For instance, it is claimed that the 'Aladdin team is responsible for running risk analyses on client portfolios and performing extensive quality control, so you can focus on the important work of analyzing and interpreting results' (BlackRock, 2020a). The human actors are envisioned to be more active when it comes to the stages of quality control and interpretation. There is a partitioning of the roles within this framing of the system. The outcome, in this type of vision, remains 'attributable' (Amoore, 2020: 154) to the human despite it being built from more automated foundations. What is clear here is that not all aspects of risk analytics and management are seen to be passed over entirely to the automated systems; a version of the human is retained in this algorithmic thinking – in the role of overseer – and so the limits of automation are not perceived to be breached.

Overall, there is a clear attempt to sketch these different actors as being part of a single and unified decision-making assemblage. They are not separated out entirely in these apparatuses; they remain interactive parts within a vision of a whole system. The vision being conjured is of unified and neatly combined agencies that shift in balance at different stages in the analytic process. As we have already heard, this is, it is claimed, 'one system, one database, one process' (BlackRock, 2020a). Aspects of the system gain legitimacy from either the human or the machine being prominent, yet it would seem that the overall attempt to gain legitimacy is based upon the notion of this being one complete system containing circumscribed agencies. As a consequence, this type of algorithmic thinking draws together components in the assemblage to work with an almost organic idea of the system – a system with a 'life cycle'.

A notion of the automated collective

As a theme, the notion of a unified system is a key part of the framing and rationale of Aladdin. In order to avoid the idea that automated systems are somehow detached from the human, Louise Amoore (2020) argues for a focus on the 'we' in the analysis of algorithms. Amoore argues that algorithms can mistakenly be understood as simply taking control away from humans. Going against the grain and ensuring the inclusion of the *we* is important, she argues, for seeing algorithmic processes in full and for interrogating their ethics. Amoore (2020: 57) focuses 'on the *we* invoked by Turing in this public debate precisely because it runs against the grain of contemporary moral panics amid machine autonomy and algorithmic decisions that appear to be beyond the control of the human'. The point here is not to avoid critical engagement with algorithms, but to make sure that the full picture may be seen. Amoore is also suggesting that the analysis is not distracted by reductive ideas of out-of-control algorithms (we will return to this in Chapter 4). In the case of Aladdin it is possible to see how a particular and engineered notion of the *we* is integrated into algorithmic thinking itself. In the case of Aladdin, it is not that the algorithmic system is envisioned as taking decisions beyond human control, but rather that there is a certain version of human input and human involvement that is presented. Rather than an overlooking of the *we*, it is a circumscription of roles that occurs in this vision of automation brought about through Aladdin. As such, it is the particular vision of *we-ness* in Aladdin that might be interrogated. In the descriptions of the Aladdin system the *we* is demarcated in particular ways and, potentially, with a notion of the limits and authority of algorithmic thinking in mind. Amoore (2020: 66) also notes that the 'human in the loop is an impossible subject who cannot come before an indeterminate and multiple *we*'. This is why it is necessary to question the images of the circumscribed human in accounts of automation, otherwise, as Amoore is pointing out, the relations of automation and agency might be missed, as might the way in which they are made up or depicted in these visions. Even though BlackRock incorporate a version of the *we* in their vision of automation, it is obviously more monological and much narrower in form than Amoore has argued that these algorithmic relations tend to be. This creates questions about how the framing of the technologies may potentially elide a much more complex set of relations within these assemblages.

It should also be added that the *we* that is associated with Aladdin is not just about the individual actor in relation to the algorithmic system; there is also a notion of a collective. One way in which this type of automation of decision-making, and the movement of discretion beyond the individual is presented, is as a part of the now familiar step towards a type of 'collective intelligence'. This is a commonly used term that has particular applications in this instance. Here the entanglement of agency is presented in terms of an assemblage that works together to improve decision-making, particularly with regard to assessments of risk. It is, for instance, suggested that 'Aladdin drives Collective Intelligence by providing tools to help your organization communicate effectively, address problems quickly, and make informed decisions at every step of the investment process' (BlackRock, 2020c). The notion of collective intelligence is an important one in this regard, with the software presented not as disrupting but as 'powering collective intelligence' (BlackRock, 2020c). A question we might ask here is about the work that is being done by the term *collective*. What is collected within it and what is not? This is not answered directly, but is rather implied. It becomes a kind of imagined collective within which the algorithm takes on a presence.

The use of the notion of a collective allows the technology to presented as being more-than-technology, picturing it instead as something that is based upon a kind of combined intelligence or a combined knowing (Chapters 4 and 5 will deal with knowing more directly). For instance, BlackRock add that 'Aladdin's Collective Intelligence gets better with every new user, and every new asset that joins the platform' (BlackRock, 2020e). The vision is of a progressing bricolage of intelligence, in which units are added to the assemblage in degrees. Each increment furthers what it knows and, in this picture, furthers the overall capacity and therefore intelligence of the system. The collective is presented as being held or pinned together by the software. The greater the scale of the collective interacting with this software, the more perfect, it would seem, is the decision. What tends to happen here though is that the exact constitution of the collective itself remains unclear – as does the notion of 'collective intelligence' (BlackRock, 2020f). The accounts of a collective lend authority by suggesting a progressive growth in the intelligence of the system. The collective becomes a notional or imagined collective of which organizations and institutions are encouraged to be a part. The *we* here is a tool for exercising a sense of authority over decisions.

The power of the collective combination of components is a driving theme. The focus shifts to what a collective mode of automated reasoning is able to achieve. Unsurprisingly, this imagined collective becomes the means by which decision-making is refined and is also the basis of a type of vision of a comprehensive form of knowing. In short, despite the collective being unspecified it is still offered as a way of developing new types of knowing. In other words, the focus moves towards the promises of that ideal. In one passage, for instance, BlackRock suggest that the 'collective intelligence' facilitated by Aladdin can help its users to, as they put it in a series of five summative properties:

See Clearer – Start every day a step ahead with a real-time view of your exposures and risks across every product and asset class, so you can make informed decisions.

Work Smarter – Aladdin handles the extensive data processing needed to support investment management, helping put the information you need to be effective at your fingertips.

Move Faster – You can move fast on opportunities with Aladdin's ability to process enormous amounts of data quickly and on a single system.

Communicate Better – Aladdin provides a common language across teams and time zones, helping to connect your people, data and processes seamlessly so everyone is always on the same page.

Scale Further – Grow your business in an efficient and controlled way using a battle-tested platform, relied on by BlackRock and other large investment managers around the world. (BlackRock, 2020)

There are echoes here of the properties of a broader 'data imaginary' that I have previously explored (see Beer, 2019a), especially in the ideals of speediness, prostheticized sight and smartness. This passage is suggestive of how aspects of decision-making are diffused into an unseen collective. The promises attached to this diffused collective are that the organization will advance on several fronts including getting quicker, growing and seeing more clearly. The balance of human and machine agency is not always clear, yet it is facilitated by these powerful promises about what might be achieved. The notion of a perfectible decision-making process underpins this set-up of the software, and it appears that the idea of a mass of combined human and technological agency is understood to be create new opportunities for visibility and speed of decision (when the division of labour is organized in ways deemed valid). It is almost as if the automated collective achieves credibility by being too complex and vast to fully comprehend. This automated collective exists as an impression. The vision of an unknown collective is mobilized to generate and legitimate knowledge and insight. Overall, what emerges is the notion of the automated collective, a notion that idealizes and legitimates processes of algorithmically informed decision-making. This automated collective contains within it ideals of a perfect type of decision-making in which particular relations between human and algorithmic agency are sketched out in a priori form. As we have already seen, within this collective the human is positioned as overseeing the industrious work of the algorithm.

Conclusion

It is important to note that this chapter has dealt directly with the framing of algorithmic systems. The implicit argument of this chapter is that such framings are important if we are to grasp the tensions of algorithmic thinking. These tensions reside within or are reflected through these framings. To analyse algorithmic systems it can be useful to think about how they are envisioned and then to reflect on both the impressions that these visions create and the agendas and objectives that are written into them. This is, of course, not the full picture. Yet at the same time such a focus provides insights into the way that algorithmic thinking operates and how it tangles with existing limits and tensions. The materials used here tell a part of this story and are intended to be suggestive of the types of ideas and ideals that are projected onto such algorithmic systems. They reveal things about those technologies and the social relations into which they are pitched, but their subtleties also reveal more about broader structural forces, sentiments and intentions. These technologies may not necessarily match up to these framings; these are, after all, examples of the myths of an algorithmic new life, yet these commercial framings are a crucial part of those technologies and their implementations. At the same time, in their visions of a new life, they reveal aspects of the underpinning tensions of algorithmic thinking. In this case, with the forces pushing towards humanlessness in mind (as discussed in Chapter 2), the aim was to think about how these envisionings retain the presence of the human.

Beyond the cases I have looked at in this chapter, there have, of course, been many other instances in which overstepping has occurred and in which the limits of algorithmic thinking have been crossed in ways that have created problems or issues. There is not always a subtle awareness expressed of what is considered to be an acceptable implementation of algorithms. The existence of limits within algorithmic thinking does not mean that they are always carefully navigated, and the concept of overstepping can be used to explore such instances as well as the navigation of those limits. Some of these oversteps become high-profile instances (for one example see Beer, 2020); there are many others that do not reach such notoriety yet might still have profound and far-reaching outcomes. Where overstepping has occurred it has led to a pushing-back against the processes of automation. Where overstepping occurs the extent to which automation is applied in the future can be limited by the fear of overstepping again. As a result, a notion of what would constitute overstepping is a part of how algorithmic thinking is framed and deployed. A notion of what would represent an overstep is a part of how the perceived limits of automation are navigated. I have only started to draw out the subtleties of this active engagement with these limits and boundaries in this chapter. Looking at the framing of a particular system such as Aladdin can reveal some of these subtleties, which are often engrained into the type of discourse used and the way that processes and outcomes are spoken of in certain terms. Continuing this analysis would require further case studies that enable the framing of automation and of algorithms to be explored in detail – it would also require such an analysis of instances where overstepping has occurred and where automation was rejected, or where it clearly unsettled social relations to the point where the limits subsequently tighten and become less elastic to future expansion.

The fear of the erasure of the human in decision-making processes hovers in the background. The expansion of algorithmic processes is in part facilitated by the avoidance of a vision that might be seen to constitute *too much automation*. And hence, there is an established and inbuilt, if not always perfectly observed, notion of overstepping. Any account that draws the limits of algorithmic thinking at the same time facilities the potential breaching and expansion of those limits (as I will discuss in Chapter 4). If a system were to be seen to be too automated, it may lead to the restriction of that system. The management of the visions of the circumscription of agency reassures and is likely to facilitate their expansion. The future of automation is based upon the navigation of these perceived limits of automation and a sensitivity to the points at which automation may be seen to be overstepping those limits. It is in the way that human and algorithmic agency are blended that a form of authenticity is sought. Knowingly, given the dystopian images it might evoke, the Aladdin case study suggests that the human is not erased but is retained in a particular role within the imagined collective.

What this indicates is that the art of *being algorithmic* is to be found in the particular visions of entangled agencies that are presented. The developments in automated asset and risk management outlined in this chapter inevitably raise questions of responsibility (Hall, 2017: 501), accountability (Neyland, 2019: 45–68) and attributability (Amoore, 2020: 154). Alex Hall (2017: 500) has pointed out that 'viewing discretion along conventional lines – as an individual judgement constrained (but also enabled) by legal or policy rules – has little purchase in burgeoning contexts of automation'. Hall adds that 'this is because it tends to replicate the strong (and problematic) separation of humans and technologies in decision making' (Hall, 2017: 500). As we have seen in this chapter, there is an attempt to try to subtly circumscribe a human presence in the case of BlackRock's Aladdin system. In other words, there is a sense that the perfectibility of decision-making is based around a set of ideals in which narrowed aspects of human discretion are maintained within the assemblage. Using the concept of *discretio* to think back across historical understandings of decision-making, Hall argues that:

> There are clearly no straightforward analogies to be drawn. What the history of *discretio* reminds us, however, is that each configuration of rule and *discretio* produces different discerning subjects. So, the real question is not whether the rule or judgement holds sway; in our times, this question is widely posed as whether algorithms or humans have power in decisions in public life. Rather, the question becomes what kind of discerning and discretionary subject – with what kind of qualities and capacities – is required by the rule and brought into being through discretion? (Hall, 2017: 501)

Hall is urging us towards an analysis of the different types of 'discretionary subject' that are brought into existence by these new algorithmic arrangements. In this particular case study of Aladdin, we see the beginning of the building of a certain type of discretionary subject that is locked into but is not necessarily controlled by the algorithmic aspects of the systems of which they are a part. The move towards *being algorithmic*, I would suggest, is not simply one in which the discretionary subject is removed but is instead where they are remade and circumscribed in ways that facilitate the expansion and spread of algorithmic processes while actively managing the tensions brought about by the potential erasure of the human. The discretionary subject observed in this chapter is one charged with a form of oversight within the collective of the algorithmic factory. What we are seeing in this case study of Aladdin is an example of how the focus on algorithmic outputs, as Amoore (2020: 166) explains, 'annul and disavow the opacity of their underlying relations in the name of crystalline clarity of the output'. Inevitably the type of commercial framings I have focused on in this chapter, which aim to give an impression of a sleek system, are likely to do just this. A notion of a circumscribed human actor is incorporated into the mix, while the opacity of the underlying relations to which Amoore is referring still remain.

Finally, the Aladdin case study and the Investec advert provide glimpses of the way that the human is retained when automated systems are envisioned. The human is placed into particular roles within what is imagined to be a unified and collective system. We only began to see moments of this in this chapter, yet those glimpses are illustrative of how decision-making processes and the category of the decision-maker are being actively remade. Selecting one widely used and influential software package in Aladdin allows this combination to be explored in one particular context; there will be many others. It is only a single case study and however influential and important this software is it can only tell us part of the story. Indeed, one of the points being made here is that we may find different accounts of the entanglement and circumscription of agency within different sectors and in different applications of algorithmic thinking. This case study enables the beginning of a picture of how these relations work so that we might examine the perceived limits of algorithmic thinking in different contexts. The perfection and legitimacy of decisions is defined in the particular combinations of agency that occur. It is here, in this envisioned blend of agencies, that automated systems and decisions appear to gain authenticity and legitimacy. How much legitimacy they gain depends on the type of balance of automation and human decision-making that is sought and how that fits with or is representative of the values of those making the judgements.

One way in which automated decision-making is rendered legitimate is through carefully defined entanglements and circumscriptions of agency. Katherine Hayles has observed that there is a problem if 'cognition is too distributed, agency is exercised through too many actors, and the interactions are too recursive and complex for any simple notions of control to obtain' (Hayles, 2017: 203). The accounts of cognition and automation built into these expanding systems seek to ensure that they do not appear to be running out of control and that their thinking isn't too distributed into the system. In short, the perceived limits of algorithmic thinking are themselves implicit in the expansion of automation. Overstepping is an intrinsic part of algorithmic thinking itself. It holds a sense of its own limits (as we will see explored further in Chapters 4 and 5). The moments when steps in automation are considered to represent *too much automation* will be moments of heightened tension that will define developments in algorithmic systems and their various integrations. Returning to where we started this chapter, the potentials of algorithms are tempered by a sense that they are inflexible and that they might come to exert too much control. Algorithmic thinking is always in tension between these forces, with the human being downgraded or removed to make space for automation on the one hand (as discussed in Chapter 2) and the desire to reinforce the place of human control, oversight and flexibility on the other (as explored in this chapter). The limits of algorithmic thinking are always likely to be sites of tension concerning just how enmeshed human agency is within these systems. There are other limits; the next one I will look at concerns the limits around what is known and what is knowable.

4

(Dreaming of) Super Cognizers and the Stretching of the Known

Impressions of *an algorithmic new life* inevitably have their boundaries. In a recent analysis of 'eight public engagement experiments', Annette Markham (2021) explored in some detail the limits that solidify within framings of algorithmic futures. Markham found that there were strong and persistent frames in place that secured certain visions of future devices, data and algorithmic social formations. Yet these visions are not simply accepted without question. The discussions at these events, Markham found, revealed that, when prompted, a critical understanding of the platforms and technologies could be readily articulated by the participants. Despite the critical ruminations of these participants, Markham also found that the framing of particular future scenarios persisted and carried with them a strong sense of inevitability. The future was somehow fixed. There was something obdurate about these framings; something that made them seem unavoidable. Indeed, Markham's (2021: 384) argument is that such 'discursive patterns continually strengthen the dominant frames of *inevitability* and *powerlessness*'. This would suggest that algorithmic thinking, as we discussed in Chapter 1, has a strong sense of the future inscribed within it. More than this though, it is also suggestive of how robust and irresistible those framings have become.

Continuing with some of the themes covered so far, this chapter explores the way that algorithmic thinking comes to have limits or boundaries that constrain social forms. Or, more specifically, it reflects on how the limits of the known and the knowable are an active part of the tensions of algorithmic thinking. To be clear from the outset, this chapter is not trying to position or establish those limits, nor is it claiming that they are fixed and secure, rather it explores the types of tensions that arise at such boundaries and how those tensions might be understood. As Markham identified, however sturdy these limits might appear there remains scope for them to shift if provoked to do so, with the possibility to imagine alternatives outside those existing frames, especially if the means are found to support and encourage different ways of thinking about the possibilities. As I will focus upon in this chapter, these sites of movement and tension are associated with what is known and what is thought to be knowable. Extending the argument from the previous chapter, I will argue here that algorithmic thinking is structured by a set of limits while also carrying a sense of how these limits might be crossed. In this particular chapter this focus on limits centres around what is known and how a sense of expanding knowability produces tensions. The very idea that the boundaries of knowing and the knowable can be crossed is, I will argue, a part of the envisioning of algorithms. The question then becomes one of whether such breaches of the knowable are within the framing of algorithmic futures or if algorithmic thinking itself is based upon a sense of its own limits being porous. In other words, it is the very idea that the limits of the known are not fixed that is defining part of algorithmic thinking (as outlined in Chapter 1). When it comes to algorithmic thinking, limits and limitlessness are both immanent.

Let us take a step back. Just over a hundred years ago, Georg Simmel (2010: 1–17: discussed in Beer, 2019b: 79–98) reflected on how the boundaries of social and individual life are experienced. Simmel explored the limits of what can be known as part of a broader exploration of life and the recasting of its parameters. Among other things, this was in part a reflection on the potential of technology to alter perspective. More than simply shifting these boundaries, Simmel's point was that as a result of technological innovation there had become a greater awareness of the boundaries of what was known, where these boundaries were drawn and the knowledge of the fact that there was more that resides beyond them. An awareness of boundaries, for Simmel, can both facilitate their crossing and, crucially, provoke a desire to cross them. He argued that a stronger sense of something existing beyond those boundaries drove their transgression. In addition to this, Simmel pointed out that where limits were previously concealed the crossing of a boundary can make those limits visible. Once outside a limit, a consciousness of the existence of that limit is far more likely. 'Our concrete, immediate life', as Simmel (2010: 3) describes it, 'lies between an upper and a lower boundary'. As life has become, he added, 'more abstract and more advanced', the result is one of a change in consciousness and the 'transcending' of boundaries. It is the crossing of the boundary that, Simmel claims, confirms 'the reality of a boundary' (Simmel, 2010: 3). In this way, shifts in perspective can potentially reveal the boundaries of what is known and can alter what is thought to be knowable.

Part of this story is how technologies – and Simmel was talking here of developments such as the telescope and microscope – facilitate a 'broadening of our sensible world' (Simmel, 2010: 3). With such developments, Simmel argued (2010: 4), perceptions of the world and life were no longer simply 'defined and limited by the natural use of the senses'. Simmel's point was anchored in the technological developments he had before him at the time and the shifts in perception that they brought about. For instance, Simmel wrote that 'since we have built eyes which see at billions of kilometres what we normally observe only at very short distances, and others which disclose the finest structures of objects at an enlargement that would have no place in our natural perception of space, this harmony has been disrupted' (Simmel, 2010: 4). Simmel claimed that an unbalancing or a deharmonization occurs once the senses are appended with technological ways of knowing and of seeing. The limits of what is known and what is not known are reconfigured by such developments. The boundaries of what can be perceived are transcended by technological developments, the result, Simmel claims, is the 'adaptation between our total organization and our world of perception' (Simmel, 2010: 4). With such developments the organization of the world is in part then formed by and comes to encompass things that are beyond human senses. We might think of the technological complexities of the last century and the extent to which the 'total organization' to which Simmel refers has happened on a massive scale. With developments in automation, questions around the extension of thinking are a clear part of this. Back then, surprisingly perhaps for someone writing over a century ago, Simmel (2010: 4) actually acknowledged the impact of these shifts on the 'structure of our cognition'. The technologies of the time were changing how people thought. According to Simmel, there is a palpable thirst for these various advancements to be pushed further, for boundaries to be breached, for perceptions to be limitless, for cognition, intelligence and thinking to be further appended and for there, in short, to be a *stretching of the known*.

One of Simmel's main observations concerned the awareness and acknowledgement of what we do not yet know. The types of technological developments he was preoccupied by, along with various cultural shifts, led to a heightened ability to think of what might not be known or what might be beyond these newly technologized senses. Simmel (2010: 5) made the point that 'we can imagine that there might be a given something in the world that we simply *cannot think of* – this represents a movement of the mental life beyond itself'. With specific developments there can, he was arguing, arise a more general change in perceptions and a reconfiguration of what is understood to be outside the known. With the altering of perspective that innovations bring, Simmel is arguing that the imagination is inspired to think beyond (or form new) boundaries and to accept that there are things that are not yet perceptible or knowable. Or, as Simmel (2010: 5) also adds, 'we ourselves know our knowing and not-knowing, and that we again know this more embracing knowledge, and its infinite potential'. This is about how limits come to be known so that they might be seen beyond. We know, Simmel explains here, what we know and also have a sense of what is not-known. In short, not only do technological developments reveal and move limits, they also bring a heightened sense of what is not known and of what is potentially beyond these newly established limits of knowledge. Simmel isn't though suggesting that these technically informed boundary breaches are straightforward or without their tensions. On this point, and as we look to apply these ideas, it should be noted that Kate Crawford (2015) has questioned whether algorithms can actually break with existing systems or break with established boundaries. Pointing exactly at the tensions that simmer, Crawford wonders if algorithms can be agonistic.

With advancing technologies and the multiple perspectives and viewpoints they might afford there is likely to have become a much more tense and frequent engagement with the limits of the known. In a recent essay on contemporary AI, with some distant echoes of Simmel's points, Paul Taylor (2021: 39) writes that 'Humans know that there is a world, that their ideas are representations of the world, and that these representations must defer to the world which will always, to some degree, elude representation'. This argument also highlights how there is a sense of there being a world that is not yet known or knowable. In this case Taylor is suggesting aspects of the unknowable fall into that category because they are somehow unrepresentable. Taylor connects this acknowledgement of the non-representable directly to developments in AI, noting that the:

> hope is that data-driven machine learning will be able to move beyond simple pattern-recognition and start to develop the organising theories about the world that seem to be an essential component of intelligence. AI software has, so far, only managed this in very constrained environments, with games such as Go or chess. (Taylor, 2021: 39)

The shift identified here is from machines excelling within the rules of the game to them actually making those rules. This shift is not just towards the protentional to see beyond boundaries but for that knowledge to be organized by these systems as well. And so we see here that there is a sense of the type of shifting perceptions and breaching of boundaries that Simmel was noting, coupled with a heightened sense of where the boundaries might be and what might be altered by new types of perception in the future. In this case, the particular move that Taylor is indicating is one in which these new types of automated systems do not just find patterns but are also capable of putting them into formalized ideas of the way things are, giving them the power to organize the bits as well as to locate them. Taking us back to Simmel's point and to a key phrase, this is very clearly and very directly an example of an *adaptation of total organization*.

As it arises here, it is worth dwelling for a moment on the question of the perception of knowing and thinking that is the frame within which this chapter is operating. Reflecting on how cognition is perceived

Katherine Hayles seeks to open up some analytical spaces that set up the possibilities for analysing exactly these issues. Let us start with one point in particular in order to set up what will follow: the notion of a spectrum of cognition. The first step Hayles takes is to think about cognition as potentially disconnected from consciousness. In a recent interview Hayles observes that 'once cognition is seen not to require consciousness, it also extends to computational media in all its forms, including networked and programmable machines' (Hayles in Amoore and Piotukh, 2019: 146). The disaggregation of cognition from consciousness creates a space in which we can then consider other forms of thinking beyond those that might narrowly be considered cognitive. This creates a space for examining algorithmic thinking in terms of perspectives of knowing. Hayles goes on to further explain what it is that can be achieved from disconnecting cognition from consciousness:

> Obviously, this view of cognition has a low threshold for something to count as cognitive, but it can also scale upward in complexity to the most sophisticated human and computational cognitive achievements. Cognition in this view exists as a spectrum rather than as a single point; it also is defined as a process rather than an entity, so it is inherently dynamic and transformative. (Hayles in Amoore and Piotukh, 2019: 146)

This places cognition onto a moving spectrum that is not limited to a single point or a particular threshold but is an unfolding and ongoing process. It is not a binary between the knowing and the unknowing. Taking this spectrum type of approach to the analysis of algorithmic thinking, this chapter asks what is happening at the limits of this spectrum and how the line of the spectrum is being drawn and redrawn as it moves into the horizon. In other words, this chapter is concerned with the conditions and tensions that are occurring as the knowledge afforded by various automated systems moves outwards, stretching what is known.

Drawing on the work of Katherine Hayles in particular, the first half of this chapter explores a series of conceptual developments around cognition and knowing. In particular, picking up on some themes that have emerged in the previous chapters, it discusses Hayles concepts of the 'cognitive assemblage' and the 'cognizer'. The objective of this first part of the chapter is to draw out some of the key issues that facilitate an analysis of the type of technological boundary breaching to which Simmel referred and how this can be understood in relation to algorithmic systems. In doing so, the chapter looks at what Lucy Suchman (2019: 36) has called the 'boundaries of robot agencies'. The second part of the chapter then seeks to build on this by looking at the role of algorithmic cognizers in shaping the limits of what is currently known. This section looks at the tensions at the boundaries of what is known and looks at the role of AI and automation in pushing at the limits of the known and the perception of what is knowable. This part of the chapter reflects upon what I will describe as the dreams of super cognizers – these are algorithmic cognizers that are beyond the limits of development and which both represent those limits and how those limits might be crossed or breached. The super cognizer, as I explore it here, is a concept through which the imagined potentials of what can be known can be examined. As I will explore, the super cognizer is a concept that captures the boundary-crossing device as it is imagined and presented. Both tangible and not too distant, the super cognizer is a concept for thinking of the bridging or gatewaying through those limits. The super cognizer is a feature of the algorithmic new life that resides just around the corner. In short, this chapter is concerned with the limits of the known contained within algorithmic thinking and how those limits distil into or solidify as certain dreams of super cognizers.

Cognitive combinations

As the previous section might indicate, there is a somewhat complex layering of issues that occur when considering not just the changes to knowing that technologies bring but also how this reshapes perspectives, limits and senses of what might come to be known. Responding to these circumstances Luciana Parisi (2019: 114) has outlined what is at stake, concluding that 'thinking about thinking involves a further level of elaboration of intelligible functions, a meta-abduction established not by a second order reflection of thinking through doing, but by the emergence of a third level of abstraction, what I called the automation of automation'. Going back to ideas about total organization and the potential organizing power of advancing automation discussed already in this chapter, Parisi identifies a third level of abstraction when considering thinking. This third level occurs where automation itself becomes automated. Here we find the limits of algorithmic thinking in action. How then to handle such a third level of abstraction? How can we approach, to adapt Parisi's phrase, this thinking about thinking? The automation of automation calls for a range of analytical responses and questions. As the terms used by Parisi indicate, this is not just automation. It is, as Parisi puts it, the automation of automation. In broad terms, Parisi's point is to do with the rule-making capacity of algorithms and the learning potential of machines, writing that: 'what is at stake here is the automation of automation: the automated generation of new algorithmic rules based on the granular analysis and multimodal logical synthesis of increasing volumes of data' (Parisi, 2019: 90). In line with Paul Taylor's point, this is where algorithms create as well as follow rules. It is this more generative capacity that leads Parisi to write of the extra level of abstraction, it is also the role of algorithms in generating rules rather than simply following those rules that are already in place.

Pre-empting such a set of questions concerning thinking about thinking, Katherine Hayles has observed that 'a discussion of transformation must necessarily involve recognition of human agencies and the recent exponential growth of non-conscious cognition in technical objects' (Hayles, 2017: 83). It is this presence of non-conscious cognition in various types of devices and systems that creates questions for thinking and that also then creates questions concerning the transformation of human agency (as discussed in Chapters 2 and 3). It should be noted that Hayles is referring to major planetary-scale changes to the ecology of intelligence, and that the analysis provided, as with that of Kate Crawford's (2021) atlas approach, is seeking to operate on that type of scale. The central thrust of Hayles' point is that to understand social transformations we need to get to grips with the combinations of cognition that are unfolding. This is to take the algorithmic thinking I have been referring to and to explore how it combines agency and cognition in different forms. The cognitive processes that lead to decisions, practices, choices and so on, are now implicated by the cognition of technical objects, as Hayles puts it.

When capturing the presence of knowing and agency within algorithmic thinking, one place to start is with the fictional representations of AI and how these depictions of automation have filtered into wider consciousness. In particular, these depictions have taken the form of dominant 'hopes' and 'fears' in 'imaginings of artificial intelligence' that are 'mediated by the notion of control' (Cave and Dihal, 2019: 74). Reflecting on some prominent representations within culture, in *My Mother Was a Computer* Hayles (2005: 242–3) argued that:

> the Terminator films to A.I. and The Matrix to Philip K. Dick's Do Androids Dream of Electric Sheep. … A central dynamic in these works is the artificial life form that refuses to accept its status as a passive object and asserts its right to become a subject capable of autonomous action, which when pitted against human agency generates the conflict on which the story turns. … If we interpret the relations of humans and intelligent machines only within this paradigm, the underlying structures of domination and control continue to dictate the terms of engagement.

We see here some of the background sources that may have fed into the persistent framing of algorithmic futures that Markham (2021) identified, which mix and match with the types of futurism and commercial logics I have discussed in Chapters 2 and 3. Hayles' point is that such dominant ideas about automation as those found in these powerful visions from film,

TV and literature can distract from the actual formations of cognition, particularly as those powerful narratives tend to centre on a narrow vision of how agency operates when cognitive machines are present.

Similarly, Asp (2019: 63) has noted that the 'notion that AI might pose such a risk goes hand in hand with the ambition to build superpowerful computers capable of fully autonomous activity'. Like Hayles, Asp also indicates that a limited focus upon only these particular types of technological visions is a limiting in terms of developing more open and inclusive analyses. More specifically, Hayles argued that popular and fictional depictions of AI are predominantly concerned with overt forms of conflict and the clashing of the human with overpowering machines – combined with the fear of out-of-control machines becoming dominant and sidelining human agency altogether (see also Heffernan, 2019). For Hayles, the important thing is not about understanding the distinction or conflict between conflicting humans and machines; it is better to think about how they combine and enmesh. Conflict will inevitably exist, yet it is not the *only* aspect. Instead, this is to think in terms of the 'intimate entanglements' (Latimer and Lopez, 2019) of humans and algorithms.

In the book *Unthought*, Hayles talks, for instance, not just of conflict but of how technologies are 'enhancing and supporting' consciousness, allowing for a focus on 'the ways in which the embodied subject is embedded and immersed in environments that function as distributed cognitive systems' (Hayles, 2017: 2). This is what Hayles has previously described as the 'cognisphere' (Hayles, 2006). Hayles' focus is upon analysing the immersion of the subject in environments where automation and intelligence are embedded within and across systems. Hayles argues that 'human subjects are no longer contained – or even defined – by the boundaries of their skins' (Hayles, 2017: 2). Such a notion of boundary dissolution is not unfamiliar, but the point that Hayles is making is that it is the distribution of cognition and agency that is the important development that further breaks down such distinctions. The outcome and the context is that, for Hayles (2017: 115), we live in 'cognitive assemblages'. The human is part of an assemblage in which thinking, understanding and knowing take place. A simple way of beginning to explore this is to think about what we know or understand of the world that isn't a product of the automated and algorithmic prioritization processes of our search engines and our social media feeds – even these mundane features of our cognitive assemblages illustrate how thinking and understanding are embedded within systems. These 'cognitive assemblages' are, Hayles (2017: 4) writes, 'complex human–technical assemblages in which cognition and decision-making powers are distributed throughout the system'. What is needed are concepts that can deal with this distribution of cognition across systems. Hayles' aim is to keep the cognitive assemblage intact while understanding how its components intersect.

Interconnected cognizers

A crucial aspect of Hayles' perspective is the extent of the interconnection of the individual actor within the cognitive assemblage. Hayles develops the notion of interconnected agencies, explaining that:

> Because humans and technical systems in a cognitive assemblage are interconnected, the cognitive decisions of each affect the others, with interactions occurring across the full range of human cognition, including consciousness/unconscious, the cognitive nonconscious, and the sensory/perceptual systems that send signals to the central nervous system. Moreover, human decisions and interpretations interact with the technical systems, sometimes decisively affecting the contexts in which they operate. (Hayles, 2017: 118)

This is a relational approach to the cognitive assemblage. Here Hayles advances this deeply interconnected and relational perspective on cognition and agency, in which decisions affect other entities within the assemblage. As Hayles (2017: 2) puts it, 'the cognitive nonconscious' is a 'term that crucially includes technical as well as human cognizers'.

As these assemblages change they take on new and different types of cognition, and the presence of algorithmic thinking is part of this picture. This, Hayles (2017: 3) adds, is about 'the spread of computational media into virtually all complex technical systems' which in turn then brings 'the pressing need to understand more clearly how their cognitive abilities interact with and interpenetrate human complex systems'. It is claimed that 'in an important sense, *these multi-level systems represent externalizations of human cognitive processes*' (Hayles, 2017: 25). Similarly, Rosi Braidotti has also written of such a 'multi-scalar rationality' (Braidotti, 2019: 44). Braidotti's image is also of the human situated and connected within this multi-scalar thinking, adding that 'the human subject is therefore only one of many forces that compose the distributed agency of an event' (Braidotti, 2019: 134; see also Chapter 2).

Shaped by externalities, in these terms, 'cognition is a process' (Hayles, 2017: 25). This perspective allows for an analysis of how changing systems intervene differently and how networked cognition meshes in different ways. In this type of assemblage, Hayles (2017: 32) observes, there are 'continual and pervasive interactions that flow through, within, and beyond the humans, nonhumans, cognizers, noncognizers, and material processes that make up our world'. There are two key points for Hayles; the first is about cognition as a process and the second is about cognition as interpenetration. Later in this chapter we will look at how such processes and interpenetrations occur through algorithmic empathy, the extension of the human and the absence of supervision. Hayles adds to this that 'technologies develop within complex ecologies' (Hayles, 2017: 33). The step that Hayles (2017: 39) directs us towards is the understanding of how 'cognizers' – a term Hayles uses to capture any presence with cognitive functioning – are understood for their varying properties and roles within this emergent ecology of the cognitive assemblage.

The cognizer as a conceptual tool

In a recent interview Hayles (in Amoore and Piotukh, 2019: 148) elaborates on this central concept, explaining that 'to be human in a cognitive assemblage means to participate in the deep symbiotic relation between biological and technical cognizers'. Further noting that this 'may be done with or without conscious awareness' (Hayles in Amoore and Piotukh, 2019: 148). When asked about the divide between cognizers and non-cognizers during the same interview, Hayles (in Amoore and Piotukh, 2019: 146) draws this tricky line by suggesting that 'the crucial features distinguishing cognizers from noncognizers are interpretation and choice (or selection)'. When it comes to interpretation and choice, Hayles continues, 'the two are entwined because without choice, there can be no interpretation, which requires at least two available options' (Hayles in Amoore and Piotukh, 2019: 146). It is this combination of interpretation and choice that Hayles uses to distinguish the cognizer from the non-cognizer. Distinctions within the assemblage can be made through other concepts, such as interpretation and choice, that then enable agents and actors to be discerned and lines between cognizers and non-cognizers to be pencilled in. This might also hint at how notions of other types, forms and advancements of cognition might be pictured, with advancing interpretation and choice being placed centrally within such visions.

Crucially, Hayles (2017: 141) claims that 'as the informational networks and feedback loops connecting us and our devices proliferate and deepen, we can no longer afford the illusion that consciousness alone steers our ships'. Even decisions about where non-human cognition is deployed is a product of the feedback loops that come from existing combinations of cognition. With the continuing emergence of networked systems we are, as Hayles (2017: 141) poignantly puts it, 'designing ourselves'. At the same time though, 'cognition is too distributed, agency is exercised through too many actors, and the interactions are too recursive and complex for any simple notions of control to obtain' (Hayles, 2017: 203). Control, Hayles argues, is too blunt an idea to cope with the many forms of cognition and combinations of agents (see also Cave and Dihal, 2019: 74). Hayles (2017: 203) argues that 'instead of control, *effective modes of intervention seek for inflection points at which systemic dynamics can be decisively transformed to send the cognitive assemblage in a different direction*' (emphasis in original). The analytical focus here is upon these 'inflection points' rather than simply upon control. This is to look at where the decision-making or cognitive processes move in a particular direction. The focus of the second half of this chapter will be upon these very inflection points.

One of the overarching principles in Hayles' position is encapsulated in the claim that 'we need frameworks that explore the ways in which the technologies interact with and transform the very terms in which ethical and moral decisions are formulated' (Hayles, 2017: 37–8). This is not just a transformation of practice and experience through technology, it is also to consider how moral frameworks and ideas themselves are transformed. As automated technologies advance and their inflection points are established, this requires us to think of them still as being a part of a cognitive assemblage, rather than lone advanced cognitive systems that are somehow separate or divest of human and other connections or interpenetrations (see also Crawford, 2021). Picking up on Hayles' concept of the cognizer, let us think further about this horizon of the unthought, the tensions that emerge at those inflection points and the stretching of the limits of the known within such cognitive assemblages.

Super cognizers at the limits of the known

As the previous discussion of Hayles' work suggests, what might be thought of as technical cognizers are already well established. There are many such devices exercising the key markers of the cognizer: interpretation and choice. There are lots of devices that are somehow thinking or engaging in non-conscious cognition. When we think of the stretching of the known the cognizer is a crucial way of unpicking the various interconnections within a cognitive assemblage that both drive and facilitate that expanded knowing. I would like to suggest that, within the cognitive assemblage to which Hayles refers, active alongside these cognizers are various dreams of what might be thought of as *super cognizers*. These are where interpretation and choice are pushed further, stretching what is known and what is knowable. These super cognizers, I suggest, are the visions of the next steps of technical cognition, the nearly-here-moves and almost-there-developments. These are the specific embodiments of an algorithmic new life in which knowing is expanding and new things are, seemingly, becoming knowable. The super cognizer is the inflection point at which there is a mixing of the advancement of automated thinking with the potential next steps that might be taken (on the future steps in AI, see Elliot, 2019: 201). The super cognizer's development is fuelled by a sense of where the boundaries might be and how they might be breached – and in some cases how they might be deliberately identified and then subverted (as argued by Dyer-Witheford et al, 2019).

An analysis of the super cognizer is also concerned with examining the pursuit of the extension of what cognizers can do and would be focused upon the active presence they play in what is known, what is considered to be knowable and in identifying what is not yet known. The rest of this chapter deals with the super cognizer as an embodiment of the cusps or thresholds in the advancement of cognizers by looking at the way they are operating at the current edges of knowledge (or the edges of knowledge in the algorithmic new life as I described it in Chapter 1). By looking at breakthrough moments it is at least in part possible to begin to see how notions of super cognizers are breaching, bridging or stretching the boundaries or limits of the known. We might explore the rationalities and logics that actively define the future directions of automation and intelligent systems by looking at how they are embodied or materialized in dreams of various super cognizers. Super cognizers are the bridges or gateways between stages in the development and deployment of algorithmic thinking. Here I would like to focus directly on the boundaries themselves, how they are being reformatted and how the perception of those limits are reshaped. In short, this exploration of the super cognizer looks at the inflection points.

The question this poses, before we continue, is where these inflection points might be located. One way to identify their location is to think about how stages in the development or advancement of algorithmic thinking are understood. If we turn back to the variety of different types of thinking technologies that Hayles pointed us towards, there is the possibility, it is argued, of separating them based upon levels or types of cognition:

> Suffice it to say the computational media are built in layers that proceed from the minimally cognitive (the basic selection between five volts or none, one or zero) up to increasingly sophisticated decision trees (subroutines nested inside routines, routines inside libraries, etc.) that can deal with highly ambiguous or conflicting information and arrive at interpretations about it. (Hayles in Amoore and Piotukh, 2019: 147)

As this would suggest and as was briefly discussed at the opening of this chapter, one way that the power of algorithmic developments has been imagined is in terms of a spectrum or a continuum running across from basic to more advanced forms. The particular ways that developments in AI have been separated into different levels of advancement has, Tobias Matzner argues, structured how we have come to think of these technologies and have been central to how developments, directions and limits have been established. Matzner (2019: 129) has noted that 'the idea of a continuum is important because those fields are structured by the question of how information technologies are better or worse than humans at certain tasks, like cognitive abilities, solving puzzles, mastering languages'. A continuum of development in AI is often mapped against human cognition and intelligence.

There is then a structuring or ordering that occurs across such a spectrum and also through this related benchmarking of cognition against human capabilities. If we take Matzner's observation, then it might be used to think about what the edges of that continuum represent and how, at these edges, there is a blurring of existing algorithmic technologies with those that are emerging. The continuum continues to spread outwards as the range of different algorithmic devices are implemented, with more embedded and rudimentary tasks taken over at the lower end and the advancing thinking capacities at the more complex end. The advancing automation end of that spectrum disappears into the distance, becoming an increasingly finely dotted line. A vanishing point of algorithmic thinking occurs. The super cognizer sits across that horizon. It is worth noting that Matzner (2019: 131) identifies where differences between humans and machines are forged and so the continuum can become unanchored from the limits of replicating human capacities. The result of this is that new limits can be created that aren't necessarily then about replicating human thinking. At the advanced end of the spectrum there is a focus on 'autonomy', which brings a range of questions of what might be thought of as 'transcendent intelligence' (Asp, 2019). Clearly this poses questions about where the limits of automation are located and how they are transcended. This idea of a moment in which humans are exceeded by AI has also been brought into question by Suchman (2019: 36), who asks whether there is actually any replication of the human in such technologies and seeks to 'destabilize the authority' of these notions by thinking about the differences in forms of intelligence. Elsewhere, Turner (2019: 29) notes that these developments are much more likely to take the form of a series of exceptions rather than one moment in which humans are surpassed, with human capacity already exceeded in some areas anyway.

Taking a similar approach to Matzner in forwarding this understanding of the stages of development in such automated thinking, elsewhere Harry Collins (2018: 74–99) has outlined 'six levels' of AI. These six levels, which are intended to capture different types of AI and to problematize the idea of a single moment in which technologies surpass humans, begin with a first level of 'engineered intelligence' (Collins, 2018: 75). This first level is a basic level of control over home devices and the like. It is worth noting that Collins is arguing that of the six levels identified it is Level 3 upwards – entitled 'symmetrical culture consumers' – that are yet to happen, or at least the fluency of interactions necessary for Level 3 are yet to be established. This highlights an existing sense of stages of development, with this stepped-spectrum also including a strong sense of what the next steps will be. This type of stepped-based approach can be used to give a sense of what is in place, but then beyond that point to imagine what might yet come – an instance of the myths of an algorithmic new life perhaps (see Chapter 1). Such an approach also breaks the spectrum into definable portions or phases of development. This portioning opens up spaces into which these next steps can be populated with dreams of super cognizers. The six levels identified by Collins run through to more advanced states with levels 5 and 6 being based upon highly refined forms of autonomy and also requiring substantial stretches of the imagination to envision or to grasp what the barriers and boundaries might be. Collins (2018: 98) points to the way the body limits possibility in the most advanced level that he outlines. Collins' discussion of these established stages in AI appears to encapsulate the inevitability and fixity of the future that Markham identified and which was discussed at the opening of this chapter.

Taking the spectrum of developments as a foundation for thinking at the limits of what is known, we can identify how knowing expands outwards into emerging possibilities on different fronts. What might become known and knowable is itself placed onto these spectra. The dreams of the super cognizer are a part of this spectrum and represent a series of directions in which the stretching of what is known and what is knowable is occurring. To elaborate further upon how dreams of the super cognizer take form at the unfolding limits of this spectrum of algorithmic thinking, I focus on three sets of related developments or inflection points at these limits. The three following focal points are spaces in which the super cognizer is active in reshaping what is known and what is considered knowable. These three components of algorithmic thinking are particularly concerned with stretching the known: empathy, extensions and unsupervision. Each of these three provide a particular set of inflection points at which the tensions of algorithmic thinking arise in striking ways. Each, in this sense, is illustrative of the way that the super cognizer creates tensions around what is knowable.

Algorithmic empathy

One area in which the limit of this spectrum of intelligence has been explored is in relation to emotions and emotional intelligence. The questions at the centre of such developments concern how emotions can be measured, tracked and responded to on one side, or how they might be replicated, imitated or performed by algorithmic systems on the other. At this inflection point what are considered very human forms of knowing are datafied and automated. This type of reading and engagement with emotions can already be found in a range of places, including children's toys, such as dolls that engage in dialogue aimed at producing a sense of friendship (Steeves, 2020) and a range of other toys or 'emotoys' that gather data to sense emotions (McStay and Rosner, 2021), through to 'empathic' forms of advertising, in which emotions, read through the 'coding' of facial expressions, are detected in some form in order to target advertising content (McStay, 2016: 9).

Andrew McStay's explorations of 'emotional AI' and 'empathic media' indicate the kind of developments that might be understood to be at the limits of these new forms of intelligence. In these cases it is an emotive set of properties that represent the advancement of algorithmic thinking. These algorithmic systems are engaging with emotions in different ways. McStay (2018: 3) explains that:

> artificial emotional intelligence is achieved by the capacity to see, read, listen, feel, classify and learn about emotional life. Slightly more detailed, this involves reading words and images, seeing and sensing facial expressions, gaze direction, gestures and voice. It also encompasses machines feeling our heart rate, body temperature, respiration and the electrical properties of our skin, among other bodily behaviours. Together, bodies and emotions have become machine readable.

The picture builds of responsive AI that are able to interpret data and infer emotions, or even be trained to have seemingly emotional responses themselves. The aim, as McStay puts it, is to make emotions 'machine readable'. Emotions are read through such technologies based upon the corporeal and other activity-related data available to them. Although, importantly, McStay (2018: 3) warns that he is not arguing that 'these systems *experience* emotions'. This is not to argue that these machines are emotional or feeling. Rather, McStay's point is that these seemingly emotional forms of intelligence are about the reading of emotions based upon data. The algorithmic system here is seeking to abstract data and use this to activate categories of emotions and emotional responses. The super cognizer here is not then about a suddenly thinking and feeling machine; it is about a machine able to coolly analyse emotions. A comparable example here would be the way that social media seek to categorize and then rank emotional attachments to particular pieces of past content in order to repackage them as memories (as described in Jacobsen and Beer, 2021). Here too, in these automated memory functions, data are used to try to read the content as potential memories and to decide which of these memories to present back in a social media or mobile phone user's notifications.

There is a further clarification that McStay makes on this point about the reading of emotions; this is to do with the power of the appearance of empathy. McStay's contention is that developments around emotional responsiveness are based in an ability to seem to understand and react to human emotion. The interest here, McStay (2018: 3) clarifies further, is in the 'ideas that the capacity to sense, classify behaviour and respond appropriately offers the *appearance of understanding*'. It is the data-informed appearance of understanding that McStay is highlighting; the focus is on the ability of the algorithmic system to seem understanding. And so the intelligence being pointed to here is based in both the ability to read emotions and in promoting the sense that these emotions have been understood. It is to use data and classification to make it look like emotions are being inferred, rather than actual emotional intelligence being in place. This is a kind of algorithmically mocked emotional intelligence. The more advanced the system, this would imply, the more difficult it might be to see that this is only the appearance of understanding. Such a system reacts so smoothly to inferred emotions as to go unnoticed and for this then to appear as genuine understanding itself. Given the importance placed on personalization, targeting and attention in the context of platforms-based media (see, for instance, Srnicek, 2016; Beer et al, 2019; Zuboff, 2019), the ability to mock emotional intelligence is a potentially powerful tool. It is clear from McStay's account that there are significant moves to develop these types of seemingly empathetic systems in various areas, from healthcare to consumption.

One property of a sort of super cognizer might then be this ability to read and appear to understand emotion. It may be found in technologies that seek to seamlessly appreciate emotional responses or to respond to emotional states in ways that pass without being really noticed – in the form of adapted playlists, recommendations, predictions, news feeds and so on. The pursuit of the super cognizer is the pursuit of an ever more convincing appearance of a machinic understanding of emotions. The direction is towards an algorithmic thinking with a type of built-in empathy, or, at least, a 'form of empathy' (McStay, 2018: 3) and the 'appearance of intimate insight' (McStay, 2018: 4). It is the intimacy that such an engagement with emotions can bring that is fitting with wider ideals of technological personalization and which becomes embodied in the super cognizer. Clearly, such an investment in the automation of emotions will be a site of tension. As a knowing emotional intelligence is automated, so the tensions of this knowing will arise.

The ability to capture data beyond human perception, to return again to Simmel's point about technological enhancement of perception, or to analyse data that are 'inaccessible to human senses', is an aspect that McStay (2018: 5) notes gives these machines 'strong cards'. Indeed, there is a dream here that the super cognizer can pick up on sense data in order to read and know emotional responses in ways that are beyond the ability of human actors. The potential then is for the impression to be built that this automated emotional intelligence can somehow identify emotions that humans may not themselves be aware of. That could be the impression at least. The steps outlined by McStay open up the emotions to analysis and to new types of personalized targeting and response (see also Davies, 2016). Or, at least, it opens up an appearance of the understanding of emotions to a range of interventions. McStay's (2018: 185) conclusion is that we are increasingly surrounded by and 'live with' devices performing such emotionally orientated tasks and with 'technologies that feel and are sensitive to human life'. There is a sense of a direction of travel here in which empathy and emotional intelligence are a focal point for the expansion of artificial forms of intelligence and algorithmic systems. This is an attempt to stretch what is known of the emotions. And McStay's (2018: 187) point is that as these 'technologies become more capable and embedded' so emotions will be known through data and automated forms of analysis. This is to algorithmically think about emotions. In turn, it is suggested that this creates a 'need to directly address the implications of living alongside systems that feel' (McStay, 2018: 187). In some instances the super cognizer is likely to appear to be feeling; it will give the impression of reading and being attentive to emotions. The ability to gather data and use this to read and create a sense of a response to those emotions is one aspect of *the myths of the algorithmic new life* with which I opened this book. The vision is of an emotionally responsive set of devices and environments that can both read feelings and also perform them.

The types of patents filed at the forefront of developments in AI, automation and algorithmic thinking are very wide ranging, but they can be drawn upon to create impressions of certain developments such as this type of algorithmic empathy. If we stay on the topic of the appearance of being responsive to emotions, patents can flesh out some of the dreams of the super cognizer as they come to be formulated. To illustrate the emotional and empathic forms of algorithmic thinking, let us pick out one example from a large volume of filed patent applications in this field. On 17 March 2020 the music streaming platform Spotify filed a patent titled 'Systems and methods for enhancing responsiveness to utterances having detectable emotion' (Bromand et al, 2020). The implication of the title is clear; it is a patent for a technology that can use what is said to detect emotion. The connections between music listening and emotions are, of course, well established and are not necessarily straightforward (see, for example, DeNora, 2000). The reasons why a music streaming service would hope to be able to seemingly understand and respond to emotions is perhaps obvious. It would allow a soundtrack to match the feelings of the listener, providing an automated form of emotional targeting. As with social media memories, this requires classifications of emotions and the placing of responses into grids. Connecting automated music delivery to inferred emotions would, it is implicitly assumed, produce the opportunity for a streaming platform to tailor, personalize and increase engagement to ever greater extents. These ambitions towards the establishment of emotional intelligence within algorithmic structures solidify in the form of the particular super cognizer described in Spotify's patent.

The long descriptions and various logistical and flow diagrams of the patent reveal some of the complexity of the proposed system. Such an appearance of the understanding of emotions requires a significant a structuring of emotions – emotions are rendered systematic and classifiable in this logic. Emphasizing the patent's clear objective to outline an empathic and feeling tendency, the word emotion appears as many as 340 times in that one patent document. The patent describes how emotions will be captured, with 'utterances' to be converted into emotions within an 'emotion index'. A typology of emotions is established through which this reading of emotional utterances will occur – forming a grid of emotions into which listeners are then placed (for a similar taxonomy of memories in social media, see Jacobsen and Beer, 2021). The patent explores how this classificatory process is deployed to try to respond to the emotions of the user. In its abstract the patent is said to outline the:

> methods, systems, and related products that provide emotion-sensitive responses to user's commands and other utterances received at an utterance-based user interface. Acknowledgments of user's utterances are adapted to the user and/or the user device, and emotions detected in the user's utterance that have been mapped from one or more emotion features extracted from the utterance. In some examples, extraction of a user's changing emotion during a sequence of interactions is used to generate a response to a user's uttered command. In some examples, emotion processing and command processing of natural utterances are performed asynchronously. (Bromand et al, 2020: 1)

This particular super cognizer seeks to be, it states, 'emotion-sensitive'. A personalized reading of emotions is pictured here, in which consumption is geared towards those readings. It is emotionally responsive to the utterances of the user and soundtracks that moment in ways that it sees as fitting. The connection is then automatically made between the soundtrack and the user emotion. However accurate or predictive that processes manages to be, this system would still be making such a connection within its own logic and its own preset terms. This type of algorithmic empathy being outlined by Spotify – which appears to flow with the circumstances but which would actually be highly codified – is based upon using sequences of interactions to process changing emotions and then to intervene. The automated reading of emotions within the consumer platform is the intended outcome.

Later in the Spotify patent it is explained that this system works through an 'emotional text-to-speech model' and that the planned system can come to label and classify when receiving 'training examples' (Bromand et al, 2020: 10). Trained using examples of speech, the system is presented as being able to learn to spot emotions. This super cognizer is said to be able to learn to infer emotions from text or speech (this can also then be positioned as part of wider developments in what has been called 'natural language processing'; see, for example IBM, 2020). The phrase 'emotion-sensitive' that is used is telling here. In this phrase emotions become something for algorithmic systems to be sensitized towards – which implies that emotions are something to be detected, extracted and mapped. And those seeming sensitivities are then to be used to shape consumption. This patent my not come into existence; whether or not this is the case, the patent remains indicative of the types of algorithmic thinking on the knowing of emotions that is in process. Whatever the specific technical capacities, it is the ability to turn what is said by the user into an emotion that can then be used to predict and recommend music to a consumer that gives us an insight into algorithmic empathy. In line with McStay's (2018) observations, here is an illustration of the sort of efforts that are in place to increase the type of feeling of 'empathic media'. The technical details aside, my main point here is that the dreams of super cognizers are embodied in such patents and are taking form in the type of technologies illustrated in this patent application and their underpinning logics and ideals. One specific feature of the super cognizer is that it might be seeking to read and provide emotional and seemingly empathetic responses. The algorithmic new life is based on this type of feeling super cognizer that stretches algorithmic thinking and knowing into the realm of emotions.

Algorithmic extensions

While reflecting on what patents might be used to reveal about the direction of algorithmic thinking, in a systematic study of Amazon's patents Alessandro Delfanti and Bronwyn Frey (2020) carefully unpick the type of automation that is being envisioned and, potentially, brought into existence. They treat the patent as providing a window onto the ongoing development of automation. The results reveal both what is happening and what represent the next steps in the automation developed by this massive corporation. Delfanti and Frey's focus on patents gives glimpses into the development of algorithmic thinking and reveals where the specialities of knowing are most likely to be extended. In short then, implicitly their exploration of Amazon's patents provides a series of insights into the specifics of super cognizers and how they embody and bridge these developments. In particular, their explorations identify forms of algorithmic extension. For instance, Delfanti and Frey observe that this use of patents offers insights into the seeming extension of human capabilities through automated systems. They explain that:

> The future automated warehouse is reflected in patents owned by
> Amazon. ... Exploiting their public nature, we strategically use patents
> to take a glimpse at technology that Amazon may one day introduce

in its fulfilment processes. The design stage is a major step through which the relationship between machines and humans materializes, and studying it unearths the political and contested nature of technology. But patents cannot be taken at face value. Some imagine spectacular innovations unlikely to materialize any time soon. (Delfanti and Frey, 2020: 3)

Clearly, as this fully acknowledges, there are methodological questions around what patents can reveal and how they should be treated within sociological accounts of the world. Not all patents become a reality and there are, of course, commercial imperatives in covering a range of potential directions for future technologies within patents. Remaining aware of such issues, they make clear how patents have value in suggesting the direction of these automated technologies and also how they can be used to reveal the type of thinking that structures and drives such developments. As is noted here, patents do not provide a straightforward vision into a definite future, yet they do provide a resource for seeing how that potential future is being imagined and, as Delfanti and Frey explain, the stages or steps through which human machine relations might move. There is an emphasis on unleashing possibilities and so the 'spectacular', as they put it, becomes a part of the visions found in these patents.

In this particular case, what Delfanti and Frey find – returning us to the discussions in Chapters 2 and 3 and thereby highlighting the interconnections of the tensions being explored in this book – is that the human is not to be erased within these plans but instead has a particular part to play in supporting the advancing automation of these spaces, infrastructures and practices. Delfanti and Frey found that 'a closer look at the thousands of patents Amazon owns and has applied for suggests that humans are not about to disappear anytime soon' (Delfanti and Frey, 2020: 4) – an illustration of *overstepping* shaping the direction of things perhaps (see Chapter 3). As discussed earlier in Chapter 3, there is a seemingly powerful circumscription of agency going on that demarcates the role of the human within the assemblage. Rather than erasing the human, what they find instead captures this type of circumscription of agency. With these patents, they observe:

Many portray a warehouse floor in which machines increase worker surveillance and work rhythms, for example, using a bracelet to capture data from and provide feedback about workers' movements. Others, such as task allocation algorithms or sensors that analyze available space on shelves, aim to segment the labor process and thus increase the corporation's ability to boost the productivity of its workforce. Visors and wearable technologies that capture data from workers' activities incorporate this knowledge into machinery and use it to rationalize the labor process in an ever more pervasive form of digital Taylorism. Patents materialize the company's desire for a technological future in which, rather than disappearing, humans extend machinery and become its living and sensing appendages. We refer to the new relationship emerging from such technologies as a form of *humanly extended automation.* (Delfanti and Frey, 2020: 4)

As living and sensing appendages, the human plays a particular and circumscribed role. It is the extension of knowing that is the crucial point here though. The activation of algorithms in the rhythms of the workplace is in keeping with Bucher's (2018: 56) observations that algorithmic agency is not simply about what is done but also about when it is applied. Delfanti and Frey finish here with their central concept of 'humanly extended automation'. The type of super cognizers found in the Amazon patents is based upon how human actors provide support for the automated systems in order for them to achieve advanced levels of knowing. Indeed, under detailed surveillance the human actor in this vision is part of the technological assemblage in which automated systems are central and are shaping what is known and how. Here, as was discussed in the introduction to this chapter, it is the machine that is being presented as providing the means to append or extend human senses and human knowing. The super cognizers found in these patents offer such extensions.

Taking this type of understanding of how the patents depict extensions of the human, Delfanti and Frey identify three forms that these extensions take. Explaining this further, they write that they:

have identified three key forms in which the continued presence of humans and the increased digitization of their work in the warehouse manifest as humanly extended automation. First, machines surveil workers and control and intensify their labor. This is based on increased machinic control over workers, decomposition and taskification of their activities, and further automation of managerial functions. Second, automation captures and datafies workers' activities and tacit knowledge to recursively improve workflows in a form of Taylorism that optimizes not just human work but machinic processes, too. Third, humanly extended automation is based on a division of labor between humans and robots in which workers are increasingly exchangeable with automated technology and intervene mostly to make up for robots' shortcomings. (Delfanti and Frey, 2020: 7)

Surveillance, codifying knowledge and enhancing divisions of labour – these are the three key extensions identified here. These three forms of extension of the human, or of the human as an extension of automation, represent a set of dreams of the super cognizer that are built into a potential near future. The first is about how human workers are overseen by active machines that are intended to monitor and govern the performance of those workers. In this sense, there is an ideal of a kind of ever-greater efficiency to be found in the automated oversight of human efforts (highlighting opposing forces in the tensions of algorithmic thinking, it is notable that this is a reverse of the human oversight of machine covered in Chapter 3). Second, Delfanti and Frey identify the capacity of these patented machines to gather data and to learn work practices in order to extend what is known about the warehouse. This is a kind of extension of the known into the patterns of working practices. Finally, there is the bypassing of the human through automation with the added point that the role of the human is to fill the gaps that the dreams of super cognizers have not yet been able to fill. The aim is to maximize automation within the division of labour and to maintain human activity within circumscribed roles or within roles that are simply yet to be automated. In the latter case the human is waiting to be replaced by future forms of automation. It is clear then why Delfanti and Frey are not arguing that the human is simply replaced by the machine but rather that, in this imagined warehouse workplace, they become extensions of the machine, being watched over, having the working environment adapt to their data and by requiring them to fill in the shortcomings of the currently patented machines. This is effectively a reversal of the human oversight of the algorithm identified in Chapter 2, thus suggesting the kind of tensions that work across the different dimensions of humanlessness and knowing.

That said, these three algorithmic extensions are not the limits to which the visions produced in Amazon's patents might be applied. These three extensions represent some broad trends among a vast array of potential applications. On this potential development of the extensions they identified, Delfanti and Frey add that:

> Patents describe many more technologies that may shape future FCs. Amazon owns patents for algorithmic systems that can reconfigure operations based on market forecasts, for example, prepackaging certain commodities or storing them in certain locations based on predictions that sales will increase or peak. Some patents describe robotic workstations that move pickers within the shelves, and others plan for automated systems that sort and transport packages. Dozens of patents for drones and robotic ground vehicles imagine an automated form of last-mile delivery. (Delfanti and Frey, 2020: 20)

A wide variety of tasks fall into the remit of these dreams of super cognizers. These are very active dreams. A certain type of imagination is at work here – an imagination that is attuned to finding gaps where automation is yet to spread or that is actively seeking to extend algorithmic thinking (this is part of what I will call *the will to automation* in Chapter 6). If we look through the previous list, there is a strong push towards automating an ever-greater number of processes and working out how to automate even more. This can be small tasks, initially, through to the types of advanced robotics and drones that are also outlined. There is one key feature to this that Delfanti and Frey identify, which is that in the case of the patents they studied, 'many seem to plan a future workplace in which human operators serve machines rather than vice versa' (Delfanti and Frey, 2020: 21). Algorithmic extensions are concerned then with taking a lead and, as discussed earlier, establishing rules. As such, they further note, these patents 'prefigure a future warehouse in which living labor operates as a new kind of appendage of machinery, making up for its shortages' (Delfanti and Frey, 2020: 21). Rather than the oversight of the algorithm outlined in Chapter 3, this is instead a form of algorithmic oversight. And so this particular dream of super cognizers is one in which the future is defined by the dominant role of algorithmic thinking and knowing in this particular logistically focused environment. This is illustrative of how the types of tensions identified in Chapter 1 can connect in different ways, with the tensions around humanlessness *and* knowing converging. In this case, the extension of a knowing workplace is also one in which the human actor appears to be sidelined or reduced to very specific supporting roles in which what is known is extended through algorithmic intervention – the warehouse, for example, is understood through data and algorithmic analyses. Plus, this balance is always in flux, with the constant push to expand the role of automated systems. In this context, as well as being overseen by the automated system the human is defined by their role as a type of placeholder in the system, temporality involved until even more advanced and complete automation arrives – returning us to Simmel once more, this is an instance of *total organization*. If we are to look elsewhere for similar types of developments, another illustration of this kind of algorithmic extension is in the use of AI in the handling of 'edge computing' (see Wu, 2020; Swabey, 2022). In edge computing automation is used to manage highly decentralized networks and to secure the edges and limits that are, as a result, more exposed (as discussed in Chapter 2). In such cases the AI is aimed at extending the network outwards to secure the edges. Here algorithmic extensions are understood through the lens of extreme forms of decentralization that takes networks outwards and as close as possible to devices. The algorithmic extension, as this suggests, is illustrative of how algorithmic thinking seeks to expand and extend outwards and into new roles and function (sometimes, as with the new vulnerabilities caused by edge computing, in response to its own presence).

The patents explored by Delfanti and Frey do not, of course, provide a fixed or full vision of the future; instead, they provide impressions of the detail and desire for a more data-heavy and automated warehouse space. The visions in the patents put in place the foundations for yet more automation processes and for the extension of algorithmic thinking. Delfanti and Frey argue that 'once incorporated into ever-evolving machinery, datafied worker activities and knowledge may provide the foundations for the automation of further processes' (Delfanti and Frey, 2020: 21). It would seem that algorithmic thinking of this type feeds off itself, creating processes and then identifying gaps yet to be filled – some of these are filled by human actors (for the time being), leaving a sense of contingency and a space for coming change upon change. A vacant space for the myths of the algorithmic new life to occupy. In this sense, we see here again the myths of the algorithmic new life etched into the diagrams and descriptions of these patents. Such expansionism is not fixed, even if it appears so. This is far from a certain set of outcomes. As Delfanti and Frey (2020: 21) reflect, 'at an even more basic level, we do not know whether the algorithmic and robotic technologies described in Amazon patents will ever materialize and enter the warehouse floor, or in which shape, although we cannot rule out the possibility that some may already be in place'. The present and future mix in quite uncertain ways in the stories told within and through patents, yet in the visions of algorithmic extensions they still provide insights into what I am calling here the dreams of cognizers and they also show how automated systems implicate what might be known while pushing at the limits of that knowing. A further question this poses is who is then watching over these expanding technologies as they learn new roles.

Algorithmic unsupervision

As this chapter has already reiterated, building upon the points in Chapter 3, notions of oversight are important in understanding the tensions of algorithmic thinking. Related to the previous two features is a third aspect of the dreams of the super cognizer, which is to do with the absence of supervision. The focus on how humans are projected as extensions of algorithmic systems is something that can be developed further through a focus on machine learning and the need for observation (which also opens up questions about what is known of the functioning and outputs of these systems, which I return to in Chapter 5). Adrian Mackenzie's exploration of machine learning is particularly useful in grasping this aspect of supervision. Mackenzie (2017: 80) points out that 'the capacity of machine learners to learn is closely linked to forms of observation that accompany and orient it'. Returning to the discussion of human oversight from Chapter 3, the role of observation is crucial in understanding the tensions that define the advancement of machine learning. The super cognizer is marked by the requirement, I would suggest, for diminishing observation and reduced supervision. It is through the observation by the system that observation by the human is seen to be less of a necessity. These systems watch over themselves (as discussed in Beer, 2019a). Rather than just watching, this observation becomes a part of the iterative processes of learning. As Mackenize puts it, these observations 'orient' machine learning. If we take an illustrative example to see how observations and supervision are thought to orient machine learning, and how a lack of supervision opens up limits of knowing, one automation provider claims that:

> Unlike supervised learning, unsupervised learning algorithms can identify existing structures in the base data, without humans having to define those first. This is very useful when we are working with unstructured data and wanting to explore patterns that we may not be aware of. Unsupervised learning is much more helpful than supervised learning when it comes to uncovering similarities, unusual events or anomalies in data sets and clustering data that is, on the surface, completely disconnected. (Pegus Digital, 2022)

The identification of the disconnected. The uncovering of the similar. The handling of the unusual or the anomaly. Here we see how it is the removal of supervision that is thought to enable different and concealed patterns in data to be identified. Without creating limits in the form of supervised guidance, this type of super cognizer is thought to be able to uncover and respond to the data themselves, without predefined limits being in place.

The type of observation or supervisions being conducted is crucial to the form and direction that machine learning might take and what it might be able to be seen to be able to achieve. Inevitably, what can be supervised and how are important in this form of algorithmic knowing. The gaps in what is observable are also an integral part of these processes. Mackenzie explains that:

> The optics of this observation of machine learners vary, but they are always partial or incomplete partly because of the dimensionality of vector space and partly because of the domains in which machine learning operates. Although the field is pragmatic in its commitment to classification and prediction, ... it distinguishes among three broadly different kinds of *learning* – supervised, unsupervised, and reinforcement – in terms of their observability. (Mackenzie, 2017: 80–1)

The optics, as it is put here, are important in defining the type of processes that are in place. These optics are also suggestive of how much observation is needed and how advancing automation might change the form of observation that is in place. Crucially, Mackenzie notes the existence of three kinds of learning that are defined by the type of observation that is occurring, of which unsupervised learning is one. Taking the first two of these three, Mackenzie (2017: 81) explains that 'supervised learning in general terms construct a model by training on some sample data (the training data) and then evaluating the models' effectiveness in classifying or predicting unseen test data whose actual values are already known'. The known data are held back to see how effective the machine is in predicting them. The effectiveness of the machine learning is itself supervised. Whereas, Mackenzie (2017: 81) continues, 'unsupervised machine learning techniques generally look for a range of well-characterized patterns in the data without any training or testing phases'. And so supervision is an important part of the type of learning that takes place. Anderson and Rosenfeld (1998: 421) draw a similar definition, pointing out that unsupervised learning is 'learning without feedback as to the correctness or incorrectness of response'. Here it is the feedback element that is removed. In their reflections on the 'dispositifs' of machine learning, Bechmann and Bowker (2019: 4) similarly note that 'in unsupervised learning data is placed in clusters or other pattern recognition outputs according to the structure in the data'. The super cognizer might then be recognizable from its advancing forms of unsupervised learning. Overall, what supervised and unsupervised learning have in common is, Mackenzie observes, that in both 'machine learners observe how a function (or functions) changes as a model transforms, partitions, or maps the data' (Mackenzie, 2017: 81). The learning is aimed at identifying such changes and finding ways to respond and to be responsive – it is just that in the case of unsupervised machine learning the starting point is less known, and the observation and feedback tend to be less significant in directing the outcomes.

Mackenzie's point goes further by reflecting on the role of observation alongside function and, also, by then thinking about how observation might lead to limitations in outcomes. Summarizing his position, Mackenzie (2017: 88) suggests 'that experimentality in machine learning consists of coupling operational and observational functions', adding that 'if operational functions move through or transform data, observational functions render the effects of those transformations visible'. It is this relation between the operation and observable functions – the making visible – that Mackenzie places at the centre of machine learning as a defining set of relations. Together these facilitate the functions of machine learning while also looking at the effects of these functions. This represents a complex set of processes in which a kind of action and then checking type of approach is built into machine learning. It is here, in the application of observation and supervision into function, that experimentation with the possibilities of algorithmic knowing occurs and where the limits of the knowable are potentially redrawn.

This, of course, creates questions about what it means for a machine to learn. Such questions are essentially questions about what is known and what is knowable. The combination of operation and observation is crucial to the type of learning that goes on and the way it can be understood and defined (this is discussed in more detail in Chapter 5). It is also crucial for thinking about the tensions at the limits of algorithmic knowing. Mackenzie (2017: 100) extends this point with the claim that:

> the power of machine learning to learn, its power to epistemologize, pivots around functions in disparate yet connected ways: the transformation of data through operational functions maps new subspaces in vector space and observational functions algorithmically superimpose new constraints – cost, loss, or objective functions – that direct an iterative process of optimization.

Maintaining a sense of the functions within an understanding of machine learning is one thing to take from this. Another is that learning occurs within these algorithmically defined constraints or limits. As the iterative processes of machine learning unfold, these algorithmic limits are established and maintain the direction and possibilities. It would seem that the learning in algorithmic thinking is constrained by its integrated observational functions. This is a kind of diagrammatic thinking; a thinking within lines. There are potential limits to this knowing, some of which, as Mackenzie is pointing us towards, may be overlooked if a notion of unsupervision is accepted. Indeed, Bechmann and Bowker have argued that, rather than working without limits, unsupervised machine learning actual relies upon existing and established classifications. Through an analysis of a Latent Dirichlet Allocation model they argue that 'a seemingly unsupervised model becomes extremely supervised due to classification work such as setting number of topics, cleaning data in a particular way with an a priori understanding of "meaningful" clusters and interpreting clusters with parent classes manually' (Bechmann and Bowker, 2019: 7). Established ideas and classifications then feed through into how unsupervised learning create outcomes; there are existing limits to what can be known that are in place. Here we see how visions or myths of an algorithmic new life might mix with the applications of a super cognizer. The image that unsupervised learning creates, as Bechmann and Bowker illustrate, is one to be explored and examined in understanding something like machine learning.

Such systematic thinking brings us back to the human and to what is knowable through this type of learning, whatever its actual limits and perceptions might be. The type of structures that are brought to unsupervised learning described by Bechmann and Bowker can be captured in other terms. It is argued by Mackenzie (2017: 100) that 'machine learning diagrammatically distributes learning in the operational human–machine formation'. Here the type of meshing of agencies occurs in diagrammatic form, with aspects of the machine learning processes being distributed across that diagrammatic space. This diagram could be understood to present those a priori limits to knowledge that even unsupervised machine learning is engaged with. Elaborating on this point Mackenzie (2017: 100) adds that:

> people look at curves for evidence of convergence, functions compress data into functions that support classification or predictions, and algorithms observe gradients or rates of error in relation to model parameters. In several senses, people and machines together move along curves.

Putting some of these specificities to one side for a moment, Mackenzie's image of knowledge moving along curves is a powerful one. Ideas and direction of the known are embodied in these curves and in the way they are followed. The curve and its waves then dictate what can be known or discovered, or they set out the effects of the operations of the system. Understanding algorithmic thinking in these diagrammatical terms and in terms of the limits of curves is a useful reference point for reflecting further on where the possibilities are forged and for exploring where the known and knowable are established and contained. These vectors are boundaries of a type that act to constrain algorithmic thinking, creating tensions around what is knowable.

The result of these curves and of this diagrammatic thinking, for Mackenzie, both reiterates the partial view of what is observable (something I will pick up and develop in Chapter 5) and the way that the limits of learning are established. These are put in the following terms:

> Every observer in this domain is partial because the humans cannot see lines or curves in the multidimensional data, the functions that underpin models such as logistic regression or linear regression can transform data in the vector space, but can't show how well they see it, and the processes of optimization only see the results of the model and its errors, not anything in its referential functioning. Omniscience (*apropos* master algorithms), whether fully supervised or completely unsupervised, is impossible here. (Mackenzie, 2017: 100)

Recalling Simmel's thoughts on the telescope, the ability of machine learning to see what the human cannot is part of this. The technological alteration of perception is based here upon the curves within diagrams that are not observed by human actors. Partial observation is occurring and the full transformations brought about by machine learning may not be visible or supervisable (as I will go on to explore in Chapter 5). Notions of optimization take on a structuring presence here too. This optimization is seen through the lens of models and errors. As such, there are then limits to machine learning, supervised or unsupervised, because of the possibilities of its underpinning properties and the way it is geared to follow curves towards notions of optimization with the diagram.

This would suggest that it is through observation that the limits of what is known and knowable within machine learning are established. Mackenzie (2017: 100) emphasizes again, and it is worth reiterating, that 'the operational power of machine learning depends on the diagrammatic and sometimes experimental relays between different practices of observing'. The tension then is likely to be in breaking from these limits of observation and supervision. The lines that are drawn into these diagrams shape what is possible from this type of learning. Put in other terms, can this machine learning find its own way and break with the limits of observation and supervision? Mackenzie's argument is that 'a function in isolation never learns'. Yet observation can potentially facilitate the breaking of patterns and the movement in different directions, potentially away from the curve. At least that seems to be Mackenzie's (2017: 102) argument when he writes that, despite what has been said:

> when watched or observed, even virtually, divergence has some chance. To the extent that machine learners relay references experimentally between things and people, mobilizing the production of statements and visibilities across different elements, divergence remains possible.

And so the potential of unsupervised learning beyond the curve remains a tantalizing presence and remains a part of how algorithmic thinking might seek to constitute a dreamlike super cognizer and its promises. Whether supervised or unsupervised, and whatever form observation might take, the question remains one of stretching or breaking the limits of what can be known or learnt. Algorithmic unsupervision provides the third set of properties operating at the edges of algorithmic thinking.

Conclusion

The super cognizer represents the limits of algorithmic thinking and is also the means by which these limits or boundaries might be breached. The super cognizer is an embodiment of an algorithmic new life. As such, the super cognizer can act as a bridge or a gateway between stages of development in algorithmic thinking, linking existing technologies and systems into those that are to come next. The super cognizer is imprinted with a strong sense of coming change. The super cognizer is active at what Katherine Hayles called, as we saw earlier in this chapter, the 'inflection points' of cognition. The dream of the super cognizer is not then about the future as such, it is more about the way that certain types of algorithmic thinking and its logics and rationalities find a form that is thought to be realizable and desirable in the present. The super cognizer is already underway and partly here.

Suchman has argued that whether we are caught in narratives of the control of human by machines or the neat control of machine by humans – which captures the two types of oversight explored in this chapter and in Chapter 3 – we are trapped in a 'universalising progress narrative that underwrites announcements of our supersession as human by machines' (Suchman, 2019: 37). As we saw earlier in the work of Hayles, the problem Suchman is identifying is with simply following such narratives and the lines of progress, advancement or technological development that they imply. Rather than following narratives of technological progress, this concept of the super cognizer is intended as a focal point for interrogating those narratives and their logic, as well as their power to shape implementation and integration. It is a concept that is intended to provide focal points that glimpse into the perception of boundaries and how these perceptions are actively reworked, redrawn and, crucially, reimagined. For instance, we looked at how such technologies are thought to be empathetic, and how a further ability to read and appear to understand emotions represents a boundary that visions of super cognizers both represent and at the same time indicate ways in which they might yet be crossed. The same applies to algorithmic extensions of human knowing and the development of algorithmic unsupervision.

The spectacular and dramatic are never all that far away, and they can find a materiality in the super cognizer. Reflecting on the potential for automated futures Mager and Katzenbach have observed that:

> evocations of possible or fantastic, desirable or dystopian futures are necessarily genuine sociopolitical processes with material consequences in the present. To make decisions in the present, we need future prospects, be they realistic or fantastical, for guidance and orientation. The future is then not only imagined, but it is also very concretely constructed, made, and unmade in different constellations and contexts. By guiding the making of things and services to come, imaginations of the future are co-producing the very future they envision. Hence, future visions are performative. (Mager and Katzenbach, 2021: 224)

Concrete future-making in the present can occur through fantastical or grand ideas of what is to come – the co-production of the present through imagined futures. This performativity of visions – the way that a possible next step impinges on the steps before – captures something of what is being explored in this chapter. Perhaps the focus in this chapter has been less on the future as such and a little more on the boundaries of the present, yet Mager and Katzenbach's points still pertain. Not least is the focus on the power of modes of thinking about future prospects. There is a making and unmaking of the future through such visions; that is the point made by Mager and Katzenbach. The dreams of super cognizers explored in this chapter are to be seen in terms of such a making and unmaking of futures within different contexts, particularly where the next stages or development are imagined to be in close proximity. The super cognizer is more immediate; it is, as with the algorithmic new life, a form of automated cognition that is just around the corner and just about reachable.

In terms of the tensions of algorithmic thinking, this chapter has explored the push towards the expansion of what is known and what is knowable. This is not an easy thing to explore without being subsumed or compelled by the powerful promises of such technologies. Dealing with these promises at a critical distance and making them part of the object of study is the important step. Yet, at the same time, it would be easy to reject them wholeheartedly without exploring their deep-seated rationalities, directions and potentials. That is what I have attempted to do here, to try to explore the forces that create tensions between the known and the knowable (Chapter 5 deals more directly with how this relates to the unknown and unknowability that these developments in algorithmic thinking create). Automation has pushed the boundaries of what is known and is seeking to continue to do so, yet it also poses some complex questions around what is knowable. Boundaries are sites of contestation of different types, and my central point is that this pushing back of the boundaries of knowledge by algorithmic and automated systems creates different types of tensions; not least, as I will go on to explore, when the discoveries exist outside human perception. The boundaries of what is known and what is knowable are stretched as perceptions of knowability change. Super cognizers are premised upon their ability to shape perspective and alter perception through the type of developments discussed in this chapter, including algorithmic empathy, algorithmic extensions and unsupervised learning. Such future moments are embodied in the very approach taken to developing forms of AI and in notions of 'superintelligence' (Turner, 2019: 28).

The spectacle of the super cognizer remains present. Reflecting on these types of boundary breaches that come with developments in AI, Werner Binder focuses upon the game-playing AlphaGo system and draws on the concept of social dramas. Binder's claim is that AI can be understood by looking at how the breaching of boundaries creates social dramas of different sorts. Binder explains that:

> The social drama of AlphaGo developed in several stages, each moving from breach to crisis, test, and resolution – albeit with varying degrees

of dramatization. Each stage of the drama was triggered by a cognitive breach: the publication of an unexpected technological breakthrough. The breach was followed by a social crisis – sometimes more, sometimes less pronounced – in which the significance of the breakthrough was debated and ultimately subjected to a reality test, namely competitive games arranged between the program and human top-players or – in the case of AlphaZero vs. Stockfish – between programs. (Binder, 2021: 189)

Within this drama, the tensions at the boundaries of automation are located in the way that boundaries are breached. According to Binder, the type and extent of those breaches plays out in the sense of crisis or drama that follows those breaches. Perhaps these dramas occur where the oncoming presence of the super cognizer becomes somehow disruptive in the tensions it creates. The focus of this chapter has been upon the stretching of the known, the projection of boundary breaches and the sense of the productive crises or dramas that they might bring. Taking evolution as a source of inspiration for thinking about robotic developments, John H. Long Jr (2019: 32) has suggested that 'complexity and uncertainty do not mean that evolution runs amok. The possibilities are bounded by mechanisms: selection, chance, history, and place.' Whether or not this is the case for advancing automated intelligence is unclear, yet it remains suggestive of the type of boundaries that might be in place and where the breaches might occur.

This takes us back to the start. I opened the chapter with Simmel's reflections on how technologies can move the organization of the world beyond the limits of perception. The dreams of super cognizers push this further and also seek to push beyond the realm of human thinking as well as exacerbating the type of sensory breaches to which Simmel was referring. The 'thinking about thinking' or the 'automation of automation' to which we saw Parisi (2019: 114) considering, is a process of direction that is continuing to unfold and that is being sketched out with the aim of stretching the spectrum or continuum of automative powers. The organizational presence of algorithmic and automated systems and new forms of intelligence are reaching towards more organizing functions rather than being about pattern recognition or the establishment solely of automated decision trees (for the context of this, see Mendon-Plasek, 2021). This chimes with Parisi's (2019: 94) observation that with 'machine learning, algorithms indeed are no longer mere instructions, but are rather performative of instructions'. The dreams of the super cognizer, across the three fronts described in this chapter, are geared towards being self-organizing – which is the very automation of automation to which Parisi refers. The self-organizing algorithmic system inevitably creates tensions around what is known. This is the thread I will pick up in the next chapter.

Before moving on though there is one final point from Parisi which takes us into the following chapter and into the verso of the known. Reflecting on what some of these developments in the organizing capacity of automation might mean, Parisi adds that:

> the automation of the intellect does not simply imply the subsumption of social values through a new rationalization of social thinking. The automation of automation instead concerns a meta-level of algorithmic function, whereby social thinking is not only organized by machines, but is algorithmically engendered through neural networked layers that eventuate new meaning of artificial thinking. The automation of automation therefore points out that the subsumption of the intellect in capital's valorization of automated cognition relies upon the social meaning of artificial thinking implied within the technoscientific descriptions of intelligence. (Parisi, 2019: 90)

The stepping-up to meta-levels means that the automation of automation to which Parisi refers has the potential to expand upon what is known while at the same time, as the next chapter will explore, generating new unknowns. Stretching the known, as we will see, is to also stretch the unknown. It is to this other dimension of this tension of algorithmic thinking – the unknown and the unknowable – that we will now turn. Where this chapter has thought about the force pushing at the expansion of automation and its possibilities for what can be known, the following chapter seeks to look further into the shadowy unknowns of these developments.

5

The Presences of Nonknowledge

Back in 1994, reflecting on the direction of rapidly advancing neural network technologies, the Nobel Prize-winning neural systems expert Leon N. Cooper mused on what the future might yet hold:

> I do have a concrete prediction. The twentieth century is the century of computers, telephones, cars, and airplanes. I think the twenty-first century will be the century of what we call intelligent machines – machines that combine the rapid processing power of the current machines with the ability to associate, to reason, to do sensible things. And I think these machines will just evolve. We'll have simple ones at first, and finally we're going to have reasoning machines. (Cooper, 1998: 94)

This juxtaposition of eras, as imagined a quarter of a century ago, hints at how the advancing computer science of the time, especially in its use of brain science as a foil, was beginning to see a future in which machines would hold escalating forms of intelligence. Where the past hundred years had been defined by machines, the coming hundred years, Cooper predicted, would be defined by how those machines were to become intelligent. With this, he imagined, would come an automated form of reasoning; an automation of the sensible. An advancing era of knowing was positioned on near the horizon, the new life of evolving machines was perceived to be just around the corner and it would bring with it automated forms of reasoning (see Chapter 1).

Cooper frames this in terms of what he imagined at the time would be a growing ability for computational reason. He also anticipated a relative and growing comfort with these forms of intelligence and what they might be used to achieve. His observation was that:

> We're comfortable with computers that enhance our logic, our memory, and we'll be comfortable with reasoning machines. We'll

interact with them. I think they will come just in time because of the kinds of problems we have to solve, these very complex problems that are beyond the capacity of our minds, probably will be solved in interaction with such machines. (Cooper, 1998: 94)

There are traces here of what I will call the unknowable and the tensions of algorithmic thinking that this unknowability might provoke. This is suggestive of the types of problems that exist, as it is put here, that reside beyond the capacity of the human mind. Clearly there are questions to be asked about the notions of comfort deployed in this passage, but this is part of how Cooper is picturing what he thought was likely to arrive in the coming century. It is imagined to be a time in which comfort in the reasoning powers of machines will grow in parallel with their advancing modes of automated reasoning. Cooper pictures few tensions here, although his mentioning of comfort would suggest that he is thinking already about how people might respond to reasoning being taken beyond their perception.

The other aspect of Cooper's vision concerns the types of boundaries of the known that I began to discuss in the previous chapter and which will become the main focus of this chapter. In this case, Cooper is imagining a future scenario – from that moment back in 1994 – in which the problems that exist beyond that capacity of human minds, as he puts it, can be resolved by machines. Or, as he continued, it is a point where the advances solve interactions between the human and the machine. Clearly there are echoes here of the earlier discussions of the role of the human in algorithmic systems (see Chapters 2 and 3); in this case though it is the way that the movement beyond the limits of what is understood to be the human mind engages also the boundaries around what is known and what is knowable. We have to consider that the type of vision of growing computational reasoning Cooper is drawing also presents questions about what is unknowable and what aspects of automation are beyond, to use Cooper's terms, the capacity of our minds. Looking back at this type of vision of the future enables an encounter with the notions and ideals that are part of the developments of algorithmic thinking. It also allows us to see how an engagement with the unknown and unknowable is an embedded part of that algorithmic thinking. Whereas the last chapter looked at the way that automation was premised upon a stretching of what is known, this chapter looks at the tensions this creates where new unknowns and new forms of unknowability are generated.

Taking a moment to reflect on his own vision, Cooper acknowledges that not everyone might share the same degree of acceptance. A sense of contestation and even tension finds its way into this picture of an automated future. He adds that:

> People will have various levels of comfort with this prospect. But I suspect that as matters evolve, we will, in the next hundred years or so, be interacting with machines that do really reason and that can be applied to very complex systems. (Cooper, 1998: 94)

We are returned here to the sense of inevitability of what is to come; we can also see here an embedded desire or will to automate (as I will discuss in Chapter 6). A growing sense of comfort with algorithmic and automated processes is a part of the expansion of the systems being imagined in Cooper's answers to the interview questions. Cooper understands this to be a part of an evolution of the technology, and so builds in a sense of direction or of inevitability about what will come. It illustrates that around 30 years ago there was already a strong sense that complex systems would come and that these coming systems would have a capability to reason beyond human capacities and beyond what is known.

A more extreme encounter with the future can be found in an interview conducted at around the same time; this time the step being imagined even goes beyond an expansion of machinic reason and reasoning. In the mid-1990s the neuro-computing and software researcher and author Robert Hecht-Nielsen (1998: 312) contended that:

> as we go forward, we're going to see the evergrowing capabilities of machines take over more and more activities that human's themselves perform. Hopefully, there'll be some rational selection of those activities, but probably not. And we will see this technology that we all know and love really become the last human technology because what will happen is that by a few hundred years from now, there won't be any human technologists. The machines will in fact be intelligent, and they will take on the role of building machines and designing machines and taking our orders and carrying them out.

The last human technology, it is provocatively claimed. The machine usurping the human arises again within these tense imaginings – Hecht-Nielsen suggests that rationality might well go out of the window when it comes to selecting which tasks to automate. The vision articulated in this passage goes further to suggest that these machines will be so advanced that they will even be developing and creating themselves, independent of human intervention. In this vision humans are bypassed and developments in knowledge and in the actual systems themselves will be led by these machines. The knowledge of the machines will be the preserve of the systems themselves. Again, this is an expansionist view (see Chapter 3) that builds the image of powerful and intelligent machines that are knowing and that are acting beyond the remits of human knowing. The ideas expressed by

Hecht-Nielsen are provocative, and are perhaps intended to be, especially as it emerges from an interview dialogue and from someone with an investment in the technologies. Yet, keeping such concerns in mind, this statement remains suggestive of the type of visions that were established among those building the foundations of neural network technologies and whose algorithmic thinking would become integrated into the advancing systems, even if such visions have not so far materialized in quite this form. We also need to recall here the context of this statement, which is more likely to have been founded in a period of heightened optimism about technology – despite the ambivalent note about the rationality of the selection of tasks in Hecht-Nielsen's statement. It was, of course, a time that had not yet been shaded by lived algorithmic experiences typical of more recent years. The all-encompassing possibilities lead Hecht-Nielsen to think of the neural network in particular as the 'last human technology'. He draws such a conclusion because, as he explains, beyond its existence technologies will know and advance themselves. It is a technology so powerful and knowing, this vision suggests, that human technologists will be redundant – their knowing will be done for them. These machines will be entirely working on the basis of unknowability – that is the aim behind them.

Writing around the turn of the twentieth century, in a piece reflecting on secrecy and knowledge, Georg Simmel (2017: 17) wrote that:

> Whatever quantities of knowing and not knowing must comingle, in order to make possible the detailed practical decision based upon confidence, will be determined by the historic epoch, the ranges of interests, and the individuals. The objectification of culture referred to above has sharply differentiated the amounts of knowing and not knowing essential as the condition of confidence.

Simmel identified a balance of knowing and not knowing that is required for confidence to be established. The things that are not-known are part of how legitimacy is established. The social recipe of knowing is, Simmel appears to suggest, an important part of understanding social relations. It is also necessary, he is arguing, to understand how the relations of knowing and not-knowing are not fixed but are contextual and relational – they adapt with the circumstances and are defined and produced by the social relations of the time. Simmel was concerned with how society and social connections are based around particular limits of knowledge, at one point in his essay he refers to this combination as 'reciprocal apprehension' (Simmel, 2017: 2). A shared understanding, a shared apprehension, of what is known and what is not is established in this reciprocal apprehension.

Simmel's point is that there is a certain amount of knowledge needed for social relations to function, but there are also limits and areas in which knowing and apprehension are beyond those limits. Simmel was highlighting the tensions that arise at the boundaries between what is known and what is not known. These are broader and longer-term questions here; what this chapter is specifically concerned with is how these boundaries function and are maintained, and also how these existing tensions are exacerbated where autonomous, intelligent and thinking machines are present. I would suggest that the tensions occurring in this balancing of the known and the not-known is a key aspect of algorithmic thinking. To provide a focal point for exploring these wide-ranging tensions around the unknown, this chapter focuses in upon one particularly important set of developments – neural networks or neural nets. Neural networks are taken as an example with the aim of illuminating the broader issues and tensions associated with algorithmic thinking. Neural networks represent a particularly influential and advanced set of established yet ongoing developments in this field. Neural networks also reveal important aspects of the wider politics of knowing associated with automation (for an early engagement with this wider politics of knowing technologies, see Thrift, 2006).

Taking us back to Lefebvre's notion of the 'new life' (with which I opened Chapter 1), through their 'regimes of recognition' (Amoore, 2020: 56; and discussed in Jacobsen, 2021) neural networks are active in producing tensions between the knowable and the unknowable. Focusing upon neural networks, this chapter explores the role of unexplainability and unknowability in AI. It looks at how unexplainability is something that is actively pursued in the advancement of neural networks. The authenticity and validity of neural networks is based upon the very presence of unknowability. Taking Georges Bataille's notion of 'nonknowledge' as a conceptual touchstone, the chapter explores how nonknowledge is essential to the ideals, development and mutation of neural networks. Examining neural networks and their integrated development of *depths* and *layers*, the chapter explores how the nonknowledge of neural networks is based in their form and in the discourse used to expound them. It looks at how the nonknowledge of neural networks is located and covered in these depths of layers upon which these systems are developed. The chapter argues that advancing algorithmic thinking and AI, such as neural networks, are based around the making of mysteries and that they function through the establishment and presences of nonknowledge.

Nonknowledge and its edges

Reflecting on the limits of what is known and what is knowable across a series of lectures and short writings dotted through the early part of the 1950s, Georges Bataille sought, in short bursts of work, to develop the concept of 'nonknowledge'. The sporadic nature of his engagement with the topic and the preliminary nature of the writings meant that these ideas were not complete or fully worked through before Bataille's death in 1962. The unfinished and work-in-progress qualities of this work are apparent in the provisional tone of these various provocations, yet there is enough here to be suggestive of what the concept might be used to achieve. Despite this contingent and slightly hesitant approach, these interventions on nonknowledge offer a useful frame of reference and raise some important questions to consider when exploring the tensions of algorithmic thinking, especially in relation to AI and neural networks, not least because they require reflection on the social role of *what is not known*. Bataille's thoughts provide some glimpses that may be used to reflect on the way that automation poses such questions anew.

On 12 January 1951, at the invitation of Jean Wahl, in what would turn out to be the first of his talks on the topic, Bataille gave a short lecture to the Collège Philosophique under the title the 'consequences of nonknowledge'. In that lecture he rather obliquely opened by questioning notions of assumed knowledge, pointing out that 'in the proposition: there was the sun and there were no humans, there was a subject without an object' (Bataille, 2001: 111). This type of pontification can, of course, be off-putting, yet it is worth persevering as it begins to hint at a useful set of underpinning questions about the status of knowledge. What is being suggested here, or perhaps restated, is that knowledge only exists where it is created and that there are fields outside this to which there is no knowledge – there are important unknowables. Returning to this point, Bataille (2001: 112) adds that 'a proposition that isn't logically doubtful, but that makes the mind uneasy, induces an imbalance: an object independent of any subject'. What Bataille seems to be identifying here is how, because of a sense of unease or imbalance, we know when we are moving into the domain of nonknowledge. Nonknowledge creates uneasiness, he is arguing, because it is a space in which we cannot be certain or sure, we sometimes become aware of that uncertainty. It is the dislocation of subject and object that Bataille uses as a starting point for elaborating on nonknowledge more generally.

This nonknowledge to which Bataille is referring has few limits; it is more of an open and unfathomable expanse of unknowing. Bataille (2001: 112) goes as far as to indicate its ubiquitous presence, adding that the 'the nonknowledge that I am talking about, in search of its consequences, is everywhere'. It may be everywhere yet it remains in tension with knowledge too. For Bataille the limitlessness of nonknowledge is disconcerting. The limits of the known make it more comfortable whereas nonknowledge is unsettling. Bataille never quite gets to grips with the core concept in these fragments, but he does draw some key aspects together, suggesting for instance that 'to specify what I mean by nonknowledge: that which results from every proposition when we are looking to go to the fundamental depths of its content ... which makes us uneasy' (Bataille, 2001: 112). There is an uneasiness that is a defining property of nonknowledge, it would seem. It is a sense of uncertainty. Nonknowledge is everywhere and it produces an uneasiness that knowledge can counter. The challenging nature of the issues that Bataille is trying to grasp here perhaps contributes to his travails. Bataille struggles around the edges of what nonknowledge might be, drawing out some points that suggest that once it is identified it becomes more difficult to define as nonknowledge. Here we can see the edges of nonknowledge. He points out, for instance, that 'Insofar as I know things obliquely, my pretension to nonknowledge is an empty pretension' (Bataille, 2001: 112). Nonknowledge is a necessary presence but where it comes on to the radar its status alters.

Given how hard the unknowable is to define, Bataille uses two particular examples to illustrate his point. He uses both death and laughter as illustrative examples intended to advance what is meant by nonknowledge and how its presence is part of the functioning of the social. A key point he makes is that despite the uneasiness it can create, nonknowledge is inevitably a part of social relations. In the cases of death and laughter, it is argued, both are not fully knowable and have aspects of nonknowledge that are an unerasable part of them. Bataille writes that 'we can picture death for ourselves. We can at the same time know that this representation is incorrect. ... Nonknowledge in particular concerning death is of the same nature as nonknowledge in general' (Bataille, 2001: 113). We know that we are encountering nonknowledge when we become aware that we cannot fully know that thing. Bataille doesn't fully elaborate this point yet he builds an image of nonknowledge as being an essential part of understanding the tensions and limits of knowledge.

In part, the concept of nonknowledge is intended as an awareness of the limits of knowledge coupled with an awareness that there is more that resides beyond those limits. It is, Bataille rather cryptically adds, 'when one knows that one knows nothing, it helps a lot; one must continue thinking in order to discover the world of someone who knows that he knows nothing' (Bataille, 2001: 114). The knowledge of knowing nothing. Nonknowledge is understood as a driver for exploration too. Here the definition collapses a little into intellectual curiosity perhaps, but it does draw attention to the potential tensions that arise at those limits. These tensions are embodied in what Bataille refers to as 'the persistent uneasiness of one who searches for knowledge' (Bataille, 2001: 115). Such a person is always immersed in the discomforting presences of the limitlessness, illusiveness and uncontainability of nonknowledge. As we will see echoing in AI, the pursuit of knowledge is premised upon an acceptance of nonknowledge. In Bataille's philosophical terms, 'every time we give up the will to know, we have the possibility of touching the world with a much greater intensity' (Bataille, 2001: 115). Perhaps, as we will see, enhanced automation is in part premised upon giving up on the will to know. The reference to *intensity* is particularly notable in this phrasing. It is suggestive of how the tensions of knowledge are both active and felt. It is also suggestive of the dynamism that can occur at the limits of algorithmic thinking, which we will see translates in the pushing and promotion of nonknowledge. The question that we will reflect on in this chapter is what happens when nonknowledge is the objective and when it is actively pursued.

Spreading these reflections on nonknowledge further over a short series of talks and writings, in a 1952 lecture Bataille thinks about the edges of nonknowledge in terms of consciousness and what people are conscious of. In this particular talk Bataille points out that 'the consciousness of nonknowledge [is] a consciousness of the absence of consciousness' (Bataille, 2001: 129). Again, the presence of nonknowledge requires an acknowledgement of the limits of what is known or what the limits of consciousness might be. Bataille concludes here that nonknowledge is 'what thought cannot conceive' (Bataille, 2001: 131). More than simply what is not known, it is also what cannot yet be conceived.

Picking this thread concerning conception up in a 1953 lecture on laughter, Bataille claims that:

> Knowledge demands a certain stability of things known. In any case, the domain of the known is, in one sense at least, a stable domain, where one recognizes oneself, where one recovers oneself, whereas in the unknown there isn't necessarily any movement, things can even be quite immobile, but there is no guarantee of stability. Stability can exist, but there is not even any guarantee as to the limits of the movements that can occur. (Bataille, 2001: 133)

The seeming stability of knowledge is the key point here. Knowledge requires a certain stability for its legitimacy, with what is known being relatively stable. Nonknowledge does not require such stability or limits, and so the possibilities remain for it to be more open and less discernible. Here Bataille seeks to use the lack of stability as a way of defining nonknowledge. Similarly, the suggestion here is that the known has limits – from which it derives stability – in ways that nonknowledge does not.

Turning again to the example of laughter, Bataille wonders further about these limits and the instability of nonknowledge:

> Suppose that the laughable is not only unknown, but unknowable. We still have to envision a possibility. The laughable could simply be the unknowable. In other words, the unknown character of the laughable would not be accidental, but essential. (Bataille, 2001: 135)

Laughter works through nonknowledge; it needs elements that cannot be conceived. Nonknowledge, Bataille argues, is an essential part of the function of this social act. Here is a contrast between the simply unknown and actually unknowable. If it were known then it might not work – more than this though, Bataille's point is that it might even be unknowable. Laughing and AI perhaps have something in common here; they both, as I will go on to discuss, require the unknown and the unknowable in order to function. Laughter works because of what cannot be known. Like laughter, we might need to reflect on where automation too is not only unknown but may also be unknowable. Like laughter, perhaps the presence of nonknowledge is understood to be essential to advancing forms of algorithmic thinking.

Nonknowledge in bits

Before reflecting on the presences of nonknowledge in algorithmic thinking, there are some further clarifications that Bataille offers that are worth dwelling upon for a moment. As mentioned, due to the contingent nature of the writings, Bataille's approach towards conceptualizing nonknowledge was perhaps a little piecemeal and as a result was not entirely systematic. In a notebook from an incomplete project, Bataille also added a further summary that provides a slightly more structured outline of what nonknowledge might constitute. Bataille begins with three key features, pointing out that:

> There are several aspects of nonknowledge:
>
> 1. Going into the world of the possible up to the point where the possible agreement fails. To refer to the possible and since the impossible is there saying to itself that the possible is coming to an end, it is as if it weren't coming to an end.
> 2. Beyond the possible, there is that which does not deceive us, as the possible obviously deceives us, since it comes to an end. Beyond the possible, I can erect what would not have the limit of the possible. But I prepare for this end in projecting into the impossible a false response to my need for an impossible possible. Meanwhile, I can tell myself that this is something that won't deceive me the way something deceives me, I ready myself for self-deception.
> 3. Within human limits, knowledge is contradicted by numerous and complex movements. (Bataille, 2001: 246)

We have to remember that these were only notes on what was an ongoing and incomplete project, yet they provide a further insight into the core properties of nonknowledge that Bataille was reflecting upon. The three previous points do not necessarily fully clarify the direction taken by Bataille, yet, as well as adding in the tensions that the contradiction of knowledge can bring, they add the issue of possibility into this conceptualization.

The possible becomes one of the limits around which nonknowledge is defined – with the movement beyond what is possible revealing the domain of nonknowledge. Here we see how the tensions of knowing are also tensions around what is perceived to be possible. The third point listed, which suggests that knowledge is itself mobile and can be contradictory, adds something to Bataille's earlier reflections by indicating that knowledge itself is not completely stable even if it has the appearance of being so. It is not then that knowledge is stable and nonknowledge is unstable, this is just how they are perceived. Rather, they both have instability within them, based upon the limits of perception and the extent of that instability. It is the presentation of being stable that demarcates knowledge and the known.

Foreshadowing the arguments I will go on to develop in this chapter, Bataille also argues that there are instances where nonknowledge is pursued, especially where it might remove the bonds and boundaries of knowledge and open up existing limits. Here nonknowledge becomes something to be achieved. It is not necessarily transformed from the unknown and into knowledge in this process, but rather the presence of nonknowledge is acknowledged, accepted and even expanded. Nonknowledge does not need to be turned into knowledge in order for it then to be part of a system. This pursuit of nonknowledge is, Bataille argues:

> because there is a desire for nonknowledge, to be delivered from the bonds of knowledge, even if knowledge is reassuring. The changes of the possible into the impossible, of knowledge into non-knowledge, from the moment when the possible is systematized, and when the impossible is no longer evasive, if the entirety of the possible subsists, do not make themselves known through the ordinary method of knowledge – to the contrary. (Bataille, 2001: 246)

The notion of a desire for nonknowledge is crucial here. The instability and limitlessness of nonknowledge are appealing. These can fuel a desire for nonknowledge. Nonknowledge has its draws and attractions; it is seen as liberating or freeing in the way it is less limited and has no confines. Nonknowledge brings possibilities. This desire for nonknowledge, perhaps in a slightly different form, could be seen to be central to emerging forms of automation, as I will go on to discuss. As we will see, to use Bataille's terms, automated and algorithmic systems don't necessarily 'make themselves known through the ordinary method of knowledge'. Yet it is the active pursuit of nonknowledge and a desire for its properties that becomes apparent within neural networks.

This point concerning the desire for nonknowledge was further illustrated in the discussion that followed Bataille's 1951 lecture. After the talk Jean Wahl asked how Bataille aspired to nonknowledge (Bataille, 2001: 117). Wahl's response to the lecture identified how Bataille was embracing nonknowledge as a productive presence. Reading this exchange back, it strikes me that the developers of neural networks and Bataille perhaps have this aspiration in common. There is an aspiration for nonknowledge and a desire to use nonknowledge as a productive presence. I don't think we need to get too enwoven in the broader philosophical questions that Bataille was leaning towards in his provocations; there is something in this incomplete concept of nonknowledge that allows us to reflect on how automation brings about both the unknown and the unknowable. It also pushes us to think about how nonknowledge can act to legitimate and drive algorithmic thinking. I will explore more directly how, in the case of neural networks, there is a pursuit or desire for nonknowledge and a kind of aspiration to use it as a productive presence.

Bataille's formulations generate questions about what nonknowledge can be, how it is formed and how it achieves a sense of instability, limitlessness and possibility. I will suggest in this chapter that neural networks are aimed at expanding nonknowledge and at cultivating new nonknowledges. More than this though, their underpinning principle is that nonknowledge is required for them to function. Nonknowledge is sometimes even an ideal to be achieved in the advancement of algorithmic thinking. As the rest of this chapter will now explore, it might even be that developments in neural networks are creating new types or new categories of nonknowledge. One area of divergence from the nonknowledge that Bataille outlined could well concern the stability of the automation of nonknowledge. One conclusion that Bataille drew was that, in its tangles with nonknowledge, 'knowledge is never anything but a precarious relationship' (Bataille, 2001: 247). This precarity is typical of the tensions of algorithmic thinking. The question then is how this precarity is managed and how nonknowledge is embraced or aspired to in the development of automated systems and particularly within the long and winding development of neural networks. As Fazi has concluded, 'above all, since they are computational, deep-learning methods involve *decision-making*' (Fazi, 2021: 60). As such, we might then argue that deep-learning methods bring to life nonknowledge and actively pursue its presence.

Neural networks and the new depths of nonknowledge

In their account of the development of neural nets, written as an introduction to an oral history project that incorporated a series of detailed interviews with 17 key figures in the field of neural networks, the neural net scientist James Anderson and the science journalist Edward Rosenfeld (Anderson and Rosenfeld, 1998) observed that the background to neural networks dates back to the 1940s. This term represents, they note, a series of attempts to 'understand the human nervous systems and to build artificial systems that act the way we do, at least a little bit' (Anderson and Rosenfeld, 1998: vii). The modes of thinking of AI and machine learning are significant to understanding their particularities. In the case of neural networks it is a case of seeking to replicate the biological. Neural networks is a field of intellectual struggle in which particular forms of knowledge and automation have continued to branch outwards. The path of neural networks is fairly long, escalating in importance in the last three decades. Stephen Grossberg (1998: 193), who worked on cognitive and neural systems, went as far as to conclude that 'we are living through part of a century-long process that has gradually led to the recent flowering of neural networks'. This type of narrative, in which neural networks were undervalued for a long time before then coming to prominence, is not uncommon in these accounts. It was a technology that found its feet over time.

The roots of the story of neural networks is quite complex and is founded in the intersection of a range of different developments. The neural network field brings together different disciplines from across computer, biological and psychological sciences. Illustrative of this, it has been suggested that 'many early developments of brain theory sprang from the roots of cybernetics' with other theorists taking 'inspiration from neuroscience, physics, electrical engineering, mathematics, and even economics' (Anderson and Rosenfeld, 1998: vii). These are accounts of the blending of different bodies of knowledge, with algorithmic thinking taking inspiration from the biological and neurological in particular. As well as reiterating this cross-pollination of ideas that are seen to be the foundations for the development of these systems, the very timing of Anderson and Rosenfeld's oral history project is also suggestive of how the 1990s represented a period in which attention was turning to the possibilities of neural networks after a fairly long period in which they were not seen to necessarily warrant much promise for longer-term development.

Despite their relatively long history, neural networks continue to be of particular and growing significance. In an overview from 2017 it was observed that in 'the past 10 years, the best-performing artificial-intelligence systems – such as the speech recognizers on smartphones or Google's latest automatic translator – have resulted from a technique called "deep learning"' (Hardesty, 2017). Deep learning is central to neural networks. This deep learning is an area in which developments continue to unfold; it has shaped and is continuing to shape the processes of automation including in everyday devices as well as in more advanced applications of classification, prediction and association. In that same overview, Larry Hardesty goes on to explain that:

> Deep learning is in fact a new name for an approach to artificial intelligence called neural networks, which have been going in and out

of fashion for more than 70 years. Neural networks were first proposed in 1944 by Warren McCullough and Walter Pitts, two University of Chicago researchers who moved to MIT in 1952 as founding members of what's sometimes called the first cognitive science department. (Hardesty, 2017)

This then is a relatively long history that is often tracked back to the early to mid-1940s and specifically to the contribution of the neurophysiologist Warren McCulloch and the logician Walter Pitts (see Lepage-Richer, 2021: 197). This winding history has seen various moments of emergence of the neural network and its associated forms of deep learning but has led to the embedding of neural networks into algorithmic thinking in broader terms.

The notion of depth being attached to neural networks is important as it is here that the layering of processes is being suggested and where the complexity and potential opacity of the neural network is implied. Nonknowledge, as I will go on to argue, is present in these depths. Dating back to 1944 the neural network's trajectory has not been linear, yet it has expanded rapidly as a mode of automation in recent years. The potential wasn't originally known, it is said, but has come to be identified, particularly as research and development has expanded into areas that require alternative modes of intelligence. The increase in computational power has fuelled this. The result is described by Mackenzie (2017: 183) as a type of 'serial reinvention' over 50 years or so. With some 'enthusiasm' in the 1960s (Collins, 2018: 102) it wasn't then until the 1980s that the 'technique' saw a resurgence before again falling away a little in the early 2000s before again gaining traction in the last decade 'fuelled largely by the increased processing power of graphics chips' (Hardesty, 2017; for more on this timeline, see also Mackenzie, 2017: 183). According to Harry Collins (2018: 102), a core idea of neural nets is that although 'initially written by a human ... once written the program lives its own life, as it were; without huge effort, exactly how the program is working can remain mysterious'. A mystery is established within these systems. Human intervention is then minimized as the algorithmic thinking takes on a greater presence. New types of automated knowing then follow.

It is important, as this would suggest, to be clear that neural networks are not a sudden or entirely recent development; instead the accounts suggest it is a technology that slowly established itself and has escalated in its developments and applications. Neural networks have been a slow burn as has the acceptance of their latent nonknowledge. As far back as 1993 the neuroscientist Walter J. Freeman (1998: 42) was reflecting on how neural networks had been around for a while, commenting that it was 'remarkable the number of applications which have emerged, the great utility of this whole approach, but I don't see neural networks as a fundamentally new kind of machine'. Rather, he adds, 'I see it more as an extension of some existing ideas' (Freeman, 1998: 42). And so, three decades ago, there is a sense of this as being an already old technology that was gaining increasing attention for the possibilities it was seen to offer rather than it emerging suddenly onto the scene.

Despite this history and illustrating how this is an umbrella term for a range of developments, there are branches to the neural network genealogy that are thought of as representing something different or novel. On what is referred to as the dynamical systems approach to neural networks, for instance, Freeman hints at a potential future direction that might bring new possibilities:

> I would say that we can only barely glimpse some of the implications of that new approach and that these really will be fundamentally different machines. They'll be unimaginably more competent at certain tasks but also maybe unreliable. ... If you're really interested in artificial intelligence and going beyond the current meaning of the term and the creation of forms of intelligence that truly have biological capabilities, this is the way to do it. And what are you going to do with them when you've got them? (Freeman, 1998: 42)

It seems it is this sense of what a biological approach towards developments in AI might bring that is seen as the site of promise in neural networks. At the same time, this is also suggestive of a form of nonknowledge and the presence of the unknown in how things might come to operate effectively. It is the sense of the unimaginable in this passage that presents the future as being a site of the development of an unknowability. The question posed and left open by Freeman concerns what would be done with such possibilities. The previous passage is illustrative of a sense of what is to come – in terms of growing levels of computational competence – and also the presentation of the unimaginability of those advancements and their potential unreliability. This is something we will return to; for the moment it is important to emphasize that Freeman's reflections situate neural networks into a longer genealogy. Reflecting on this genealogy and the variation of developments falling under this umbrella term, it is worth noting that in other instances it has even been questioned if neural networks is the even right term to be using to label such systems. It has been suggested, for instance, that 'much of the work could have been done in identical form under another name' (Kohonen, 1998: 153). Any reflection on this would require a sense of how the term neural networks evolved and what was at stake as it became established in the discourses of algorithmic thinking and as the connotations of such a term became active within these movements.

It is perhaps unsurprising then that as neural networks developed there persisted issues around their definition and functions. Théo Lepage-Richer (2021: 199) notes that distinctions are an issue here and that there is 'an ambiguity regarding neural network's status as a theoretical or functional model – an ambiguity which, given McCulloch and Pitts' experimental ambitions, might be as old as the concept itself'. This point about the potentially ambiguity of the terminology hints that nonknowledge may have been implicit from the outset. Crucially, Lepage-Richer (2021: 199) adds that 'neural networks are conceived as models of knowledge acquisition based on the operationalization of what lies beyond the limits of knowledge'. Establishing the limits of knowledge, the neural network is aimed at working beyond what is known. The neural network seeks to work beyond the limits of the knowable. Despite the various ways in which neural networks might be conceptualized, this particular property is identified by Lepage-Richer as a uniting factor. It is claimed that 'the core property that unites the different systems encompassed by the concept of neural networks appears to not so much be their structural similarities, but rather their shared conceptualization of the unknown as something that can be contained and operationalized by networks of interconnected units' (Lepage-Richer, 2021: 199). This then is a more stable notion of the unknown, yet nonknowledges clearly have a presence here. Neural networks don't necessarily share the same structural or material properties, but they do seek to operate beyond the limits of knowledge and, therefore, within the unknown. This conjures images of something spectacular; at the same time we might infer that this can also take quite rudimentary even banal forms too, with Mark Andrejevic (2020: 31) noting that 'the former Google engineer who oversaw the neural networks that learned to recognize cat photos admitted that he didn't know how the system worked'. The system then is operating without a full knowledge of how it reaches its outputs. In this example there is a consciousness of the absence of consciousness, to return to Bataille's point. The engineer knew that he didn't know how the system was working. The description of that engineer is suggestive of how the neural network is not something to be understood but is a system that is left to its devices. Andrejevic (2020: 31) concludes that in general terms 'the culminating promise of automated media is the advent of artificial intelligence that can generate useful but otherwise inaccessible and incomprehensible insights'. This nonknowledge is not a by-product so much as it is a central and defining aim. They run on nonknowledge and are intended to reside beyond the explainable. Indeed, the aim is to be unexplainable. The presence of nonknowledge even seems to promote a sense of the authenticity of the system.

There are two further related properties of neural networks that have been a part of their development and which have become implicit in the ideals encased within their framings. On the one hand is self-organization and on the other is adaptation. Both in different ways require aspects of nonknowledge to function within the system. To explore these two features, let us turn briefly to two illustrative moments. In a 1993 interview the neural network scientist Teuvo Kohonen stated that:

> It is my dream that the self-organizing mapper could be used as a monitoring panel for any machine where you have to monitor dynamic variables – or if not dynamic, then with lots of parallel variables. You could have that in every airplane, jet plane, or every nuclear power station, or every car. You could see immediately what condition the system is in. That would be something like what our nervous system is doing instinctively. (Kohonen, 1998: 160)

This is indicative of an embedded impulse towards self-organizing systems operating and observing their own processes. This type of algorithmic thinking and its will to automate is based upon an ideal in which systems can organize themselves and can therefore become dynamic and responsive (as discussed in Chapter 4). Again, the biological becomes part of how this is imagined and how the functioning of self-organization might be envisioned. In this case the self-organizing system is imagined to be like a nervous system – bringing with it the nonknowledges of such a complex system. The suggestion here is that the ability to organize without human intervention is a desired outcome (illustrating the interconnections of these tensions, the posthuman securities of Chapter 2 are recalled here as are the discussions of the organizing systems from Chapter 4).

Similarly, adaptation and the ability to create systems that are able to adapt to their circumstances is another aspect of this push towards automation. In a move that recalls Bataille's point about the consciousness of the absence of consciousness, neurocomputing engineer Carver Mead in 1994 evokes the image of a 'cognitive iceberg'. In this iceberg model, as he explains it, we only know the tip of consciousness. This is the type of notion of cognition that then shapes algorithmic thinking. As such, Mead (1998: 135) reflects on how 'we get these fully formed concepts and percepts that come floating up from below'. Taking this vision of consciousness as the focal point, Mead dreams of a system that is able to conjure concepts and can use these concepts to adapt its understanding – this is reminiscent of the concepts coming from the depths of Bataille's nonknowledge. As with self-organization, it is the adaptive system that is the vision here. When asked what neural networks might offer, tellingly Mead responded 'Adaptation', adding that:

> it's the whole game. I really believe that. It's not just because of the technology I work in; it's because of the nature of the real world. It's too variable to do anything absolute. … You have to develop a higher

> level of abstraction ... and you do that by comparing things and adapting to things. The nervous system figured that out a long time ago. (Mead, 1998: 135)

Adaptation is the crucial focus for Mead. It is, as he puts it, the whole game. The reason for this is not just that these systems can allow concepts to float to the surface of their consciousness, replacing how he sees human cognition, it is also because these systems can then be active and responsive when applied in 'real world' settings – elsewhere it is uncertainty that is discussed as representing a similar problem once the technologies are moved outside an 'artificial environment' (Sejnowski, 1998: 331). Drawing a similar conclusion on the future direction of neural networks, Stephen Grossberg also argued that adaptation was going to be an important step. For Grossberg (1998: 195), the longer-term revolution he was picturing 'is about how biological measurement and control systems are designed to adapt quickly and stably in real time to a rapidly fluctuating world'. Grossberg adds that achieving this is about, he claimed, 'discovering new heuristics and mathematics with which to explain how nonlinear feedback systems can accomplish this goal'. Again notions of the biological mix into notions of nonlinear forms of knowing. The ability to work without guidance in order to be able to adapt is the central thrust of these ideas. The system is essentially aimed at being engineered to respond to unknown circumstances.

In terms of this future adaptability of neural networks, the problem Mead encountered in realizing this aim was that, as he goes on to note, 'adaptive circuits have been much harder to build robustly than I had any idea' (Mead, 1998: 137). Adaptation is sustained as an aim and reveals an underpinning principle for unexplained thinking. The vagaries of the real world, in these terms, can potentially then be matched by the ability of the system to adapt and to call up from the depths of cognition the concepts it needs to respond. This is a nonknowledge based on forms of algorithmic thinking in which self-organizing and adaptive systems react and create rules based upon unknown circumstances.

Inevitably, care needs to be taken over some of the terminology. Working around the different definitions and accounts of the terminology would be an interesting project in itself, especially as the terminology and labels are sites of tension. For the moment, it is worth simply acknowledging that the terms being used may not always have a fixed set of properties. For instance, Beatrice Fazi (2021: 58) notes that:

> Deep learning is itself a remarkably multifaceted technique. To simplify, an artificial neural-network system relies on layers of artificial neurons to process information. These layers of artificial neurons are connected and influence each other in a complex web of interacting

units, somewhat like biological neurons are understood to do in a biological brain. A lower layer of neurons performs computation and transmits this result to the layer above, enriching the final outcome of the layer at the top.

The *deep* in the phrase deep learning refers to this layering within neural networks. This production of depths through layering is something I'll return to in a moment. For now, it is the form of the neural networks that I want to focus upon. The account here brings in, again, the salience of brain and biological metaphors in the understanding of these systems and the learning facilitated by the interaction of components. This begins to flesh out the ideas and features associated with neural networks and deep learning. Even here there is a sense of the unexplainable, with Anderson (1998: 259–60) commenting that through research on the brain 'our major discovery seems to be an awareness that we really don't know what is going on'. The development of neural networks then may be based on the brain, but this is a version of the brain that was itself not fully known.

The descriptions of neural networks vary yet, as we have already seen, there remains some common features. Some of this is based in the way the system arrives at its own answers, often outside of human interaction or, sometimes, without full understanding of the outputs. This is illustrated by the following commentary on neural networks, which points out that:

> they are modelled on a theory about how the human brain operates, passing data through layers of artificial neurons until an identifiable pattern emerges. Unlike the logic circuits employed in a traditional software program, there is no way of tracking this process to identify exactly why a computer comes up with a particular answer. (Waters, 2018)

The idea of the relation between AI and the operation of the human brain is relatively widely referenced in discussions of neural networks, as has already been illustrated. There is more here though; in particular, the idea that the outcomes cannot be fully followed through the system. There is no way, it is claimed, of tracking processes. In this case Waters returns us to the point about the difficulty of understanding decisions. The complexity means that it becomes hard to grasp how an outcome was arrived at. These are algorithmic systems that seek to generate nonknowledge and to be unexplainable.

On these points of unexplainability and the non-trackability of processes, it can be useful to turn to concepts of machine learning within neural networks. Hardesty's account for instance, suggests that:

> Neural nets are a means of doing machine learning, in which a computer learns to perform some task by analyzing training examples.

Usually, the examples have been hand-labelled in advance. An object recognition system, for instance, might be fed thousands of labeled images of cars, houses, coffee cups, and so on, and it would find visual patterns in the images that consistently correlate with particular labels. (Hardesty, 2017)

The image here is of the neural network being fed information and learning from it (taking us back to the learning super cognizer from Chapter 4). The human labelling of images feeds into the teaching of these systems (for a discussion of the politics and prejudices of this processes, see Chun, 2021). Having learned from these inputs, the system is intended to adapt to those data. As the information is fed in, the neural networks operate to find patterns. Again, to understand this, the image of the human brain is evoked, if a little reticently, to help to explain how this type of learning occurs:

Modelled loosely on the human brain, a neural net consists of thousands or even millions of simple processing nodes that are densely interconnected. Most of today's neural nets are organized into layers of nodes, and they're 'feed-forward,' meaning that data moves through them in only one direction. An individual node might be connected to several nodes in the layer beneath it, from which it receives data, and several nodes in the layer above it, to which it sends data. (Hardesty, 2017)

It would seem that complex systems emerge from the networking of many nodes within these layers. The crucial aspect of this is the presence of layering in these systems. As these systems advance so the neural networks gain layers, with extra 'layers of neurons in modern versions of neural nets' (Collins, 2018: 10). The picture is of data passing across networks and also between layers within those networks. The image is volumetric and three dimensional, of data working through, between and across networks in these machine learning processes. More layers means a greater level of depth to the systems and a greater level of mysteriousness. Indeed, Fazi (2021: 61) points out, crucially, that '*Explainability* is a key word for present and future algorithmic cultures, raising equally unique social and ethical challenges'. This is to highlight more directly the importance of explainability in understanding developments in deep learning and neural networks (see also Hayles in Amoore and Piotukh, 2019: 151). The layering of networks poses ethical questions about whether they can then be explained. More than this though, the politics may well be in the ambition to make these systems unknowable in the first place.

There are depths to the nonknowledge of neural networks. When accounting for the focus on memory and how the biological might be replicated, in a story that takes the lineage back to the late 1950s, Leon Cooper described how the system they were working on 'learned using only local information available at synaptic junctions between neurons. It did not give very accurate recall, but it did work and operated as a simple kind of associative memory' (Cooper, 1998: 81). They sought to develop memory as a part of these systems, with recall at the forefront. Crucially, Cooper also adds that 'people regarded this as among the deep mysteries' (Cooper, 1998: 81). A deep mystery. This is a phrase that deserves interrogation. It pushes us to ask what these mysteries are and to think about the presences of nonknowledge in the emergence of algorithmic thinking. The importance nonknowledge here is in the functioning of the system, in what the unknown may represent and also in the active cultivation of that nonknowledge. As the system's learning gets deeper, adding more layers, so did the depth of the mystery.

Neural networks, hidden layers and the mysteries in the depths

In an interview with the *Financial Times*, Illah Nourbakhsh, a professor of robotics, observed that 'that's the odd irony of AI – the best systems happen to be the ones that are least explainable today' (Waters, 2018). From a different perspective but drawing a similar conclusion, in a recent interview Katherine Hayles (in Amoore and Piotukh, 2019: 151) has also noted that 'recursive architectures limits how much we can know about the system, a result relevant to the "hidden layer" in neural net and deep learning algorithms'. This hidden layer is an embodiment of the tensions around knowing and unknowability. These points chime with the previous discussions, which take us to the vision of the layering of neural networks that creates the depths within which processing occurs. It is the addition of depth that offers spaces for nonknowledge to be established within these systems. Nonknowledge resides in the shadows of these layered depths. The idea of the training of the neural network is often part of this, with multiple layers giving greater scope for such training. Hardesty explains that:

> When a neural net is being trained, all of its weights and thresholds are initially set to random values. Training data is fed to the bottom layer – the input layer – and it passes through the succeeding layers, getting multiplied and added together in complex ways, until it finally arrives, radically transformed, at the output layer. During training, the weights and thresholds are continually adjusted until training data with the same labels consistently yield similar outputs. (Hardesty, 2017)

This is a brief account of the way a neural net is trained, yet it still gives an impression of the type of depthfulness encountered already in this chapter.

Training requires layers and depth. The iterative processes described here are the basis of machine learning. It is a vision of the training of the technology until it reaches the required outputs, with the layers allowing complex and multiple processes to be established. The processing moves through these multiple layers, travelling further from the input. The complexity and unknowing builds due to the number of layers within these depths.

Advances in gaming have played a particularly important role in extending and enhancing these depths. Gaming developments have been built around advancing graphics processing units (GPUs) with, it has been observed, 'the architecture of a GPU [being] remarkably like that of a neural net' (Hardesty, 2017). The capacity to train these systems increased as these particular gaming systems developed. These developments are based around the depths of the layering occurring within them. The specific advancement of GPUs was among a broader range of advances that led to rapid increases in depths and layering in neural network technologies. In terms of these growing depths, Hardesty has noted that:

> Modern GPUs enabled the one-layer networks of the 1960s and the two- to three-layer networks of the 1980s to blossom into the 10-, 15-, even 50-layer networks of today. That's what the 'deep' in 'deep learning' refers to – the depth of the network's layers. And currently, deep learning is responsible for the best-performing systems in almost every area of artificial-intelligence research. (Hardesty, 2017)

The layers within the neural networks rapidly multiplied. Neural networks got deeper as the layers were added. The systems advanced in terms of their processing as the layers increased and the depth was extended. The notion of depth in deep learning is important and instructive here. These, I would suggest, are the depths of nonknowledge.

Or, perhaps a better way of putting this is to say that the layering within these systems creates spaces for nonknowledge to be cultivated. It is this adding of layers that produces the deep learning imagery that is often evoked and which also imbricates the processes to the point at which it becomes impossible to unpick. The increase in depthfulness is associated with notions of advancing AI and enhanced machine learning capacities. Depth is the root of notions of authenticity in machine learning. It is also where the greater depth leads to a greater sense of the unexplainability of the outputs – and so authenticity and unexplainability become connected. The move is towards the pursuit of such unexplainability. In other words, it is in the depths of the layered networks that mystery spaces are being brought into effect and where nonknowledge is being extended within such algorithmic thinking.

The depths proceed beyond this, adding further layers of nonknowledge. The production and pursuit of nonknowledge doesn't end there. One further aspect of the layering in neural networks, as mentioned in the chapter introduction, is what is referred to as the hidden unit or hidden layer (for an example of the ongoing discussion of how many hidden layers to include in a neural network, see Keim, 2020 or Shen et al, 2021). Reflecting on the inception of this term, the computer scientist Geoffrey Hinton (1998: 379) reveals that this phrasing was the result of a chance occurrence during the mid-1980s:

> One very lucky thing happened to me at CMU [Carnegie Mellon]. Early on when I was there, I had a graduate student called Peter Brown, who knew a lot about speech recognition and knew all about hidden Markov models. He told me about hidden Markov models while I was doing Boltzmann machines. I wish I'd understood more about them then because I only very, very slowly really understood them. ... The reason hidden units in neural nets are called hidden units is that Peter Brown told me about hidden Markov models. I decided 'hidden' was a good name for those extra units, so that's where the name 'hidden' comes from.

The development of the terminology emerged through this associated reference point. As such, the term 'hidden' in neural networks is not as straightforward or descriptive as its usage might suggest. It would seem that we cannot necessarily take the notion of the hidden layer at face value, given that the term was produced by Hinton's association between models. Yet it is still suggestive of something, which is the presence of a type of nonknowledge in how these systems were being formulated. Hinton's professed limited understanding of the models from which he derived the term also offers some uncertainty about the aspect of the hidden that was taken from the original source. Yet, at the same time, it is clearly indicative of how these layers were understood and also, perhaps, of the type of embedded objectives to produce a vast layering of processes in these systems. The transposition of the term hidden is revealing of the ideals associated with the incorporation of the unknowable. In itself, we have here both the hidden layer along with the pursuit of the integration of mystery into the system's structures. It is also suggestive of the presence of nonknowledge and of unexplainable within these emerging systems.

Depths and learning are inextricably connected in neural networks, including in this notion of the hidden layer. Indeed, the very notion of learning comes from the layering of depths – with more advanced systems layering multiple hidden layers. Learning arises from these depths and so then it becomes more difficult to explain how decisions are arrived at. According to Fazi:

A neural network is said to learn, then, because it can tweak its calculations and modify its interactions by tuning parameters via activation and back-propagation among layers until the desired output (i.e. the desired final representation) is produced. The network, however, is called deep if its structure encompasses intermediary 'hidden' layers between the input and the output. The architecture of a deep-learning system differs from that of a standard artificial neural network precisely because of the presence of these multiple non-linear hidden layers. (Fazi, 2021: 58)

It is this building up of layers that facilitates the learning capacity and which, in so doing, puts that decision deeper into the system. These hidden layers are less fathomable – so they facilitate greater depths of learning while also dictating the extent or depths of knowing and understanding of that system (see also MacKenzie, 2017: 192). In order to learn, these systems appear to require an increased level of unknowing about how they operate. Deep learning is defined by the presence of more layers, giving it that depth. This is why Fazi has argued that explainability is an important part of responding to these systems. It could be said that the consequence of the layering of deep learning is form of murkiness. Fazi (2021: 59) writes that 'because of how a deep neural network operates, relying on hidden neural layers sandwiched between the first layer of neurons (the input layer) and the last layer (the output layer), deep-learning techniques are often opaque or illegible even to the programmers that originally set them up'. This would suggest that the pursuit of knowing appears to require an unknowingness within the system, or, as I have argued, it requires the presence of nonknowledge. The greater the layering the harder these systems become to explain. Fazi goes on to expand upon this point concerning explainability, adding that 'once a deep neural network is trained (or self-trained, as in the case of AlphaGo Zero), it can be extremely difficult to explain why it gives a particular response to some data inputs and how a result has been calculated'. The decisions of the trained-up neural network become hard to trace. The trained or self-training neural network becomes harder to explain as do its outputs – this, it would seem, is an intended and perhaps even desired outcome.

Looking back, we can actually find further illustration of this problem of explainability and the presence of nonknowledge even in the early development of neural networks. The mathematician and theoretical biologist Jack D. Cowan has described visiting Wilfred Taylor to discuss his work on analogue neurons in 1956. Cowan (1998: 107) found a baffling machine in action and recollects that:

I tried to get him to explain to me how it worked. I couldn't. He didn't really understand himself what was going on, I think, and to this day in published papers you can't quite understand how it works. The learning rule he's got is not associative, and yet the performance is, for there's something fishy about it. It was all done with analog circuitry. It took him several years to build and to play with it.

The phrase 'something fishy' in this passage is perhaps suggestive of a presence of nonknowledge even in these foundational moments in neural networks from over 60 years ago. What might be thought of as more rudimentary systems, such as the one described in this passage, already had an unknowability coded into them. It seems that for a substantial time, forms of nonknowledge have been tied up with the functioning of neural-based systems. The outputs seemed to make sense in Cowan's recollections but it was how that outcome was arrived at that remained a mystery. Occurring in the depths of that analogue circuitry the unexplainable and the unknowable were already a part of the systems themselves.

In another similar instance, reflecting on some collaborative work from the late 1970s, David Rumelhart (1998: 275), who had a background in psychology and was collaborating on neural networks, remembered the:

hours and hours and hours of tinkering on the computer. We sat down and did all this in the computer and built these computer models, and we just didn't understand them. We didn't understand why they worked or why they didn't work or what was critical about them. (Rumelhart, 1998: 275)

The system was operating but they had to try to work backwards through the steps to try to understand it. These systems were created but could not be fully explained by those working on them. This is an account of working with the systems in a hands-on way to try to get them to function while at the same time not fully understanding the operations of what has been built. Further suggestion, perhaps, of the presence of nonknowledges even from the early stages of neural networks.

Despite the efforts of these researchers to understand the systems that were being created, largely so as to expand them and develop different functions, it is possible to argue that a kind of pursuit of unexplainability is inherent in such advancing forms of automation. This is algorithmic thinking with nonknowledge built into its depths. This inevitably creates tensions around understanding and explanation. In short, these programmers are creating something that potentially cannot be fully known; indeed, that is sometimes even the aim. Starkly, Fazi (2021: 59) argues that:

the automated learning choices of a deep neural network are not yet fully understood by programmers. The knowledge generated in these models remains, in part, implicit due to the non-linear nature of deep learning, its compressed information, and the distributed character of the network's representations, which rely on the many configurations of its large sets of variables. Such a complex, layered architecture entails difficulty in analytically comprehending what nodes and layers have learned and how they have interacted to transform a representation at one level into another representation at a higher, more abstract step.

This is a crucial intervention in which the issues around understanding and explainability are made clear. It is interesting that Fazi writes of the choices being made in deep neural networks as not *yet* being fully understood by programmers. The use of the word *yet* suggests that these choices can be fully known or that there is an intention to understand and explain them. The trend may possibly be in the other direction, with the aim to move those choices further into the unknown as a part of making these systems more autonomous. It is perhaps the move towards advancing the presence of nonknowledge that is the direction in which these systems continue to head. The non-linear, multi-layered and distributed forms of these systems that Fazi is describing is a step in a particular direction that is likely to move further towards the unexplainable. Perhaps, given the focus on the problem of explanation, this is actually Fazi's point: the systems are moving away from being explainable in their very form and direction of development. If we add to this the necessity for and defining presences of nonknowledge, and requirement of the presence of nonknowledge for these systems to function and gain authenticity, then the scale of the unexplainability and unknowability of such algorithmic systems is likely to be exacerbated.

This presence of nonknowledge in algorithmic thinking is not an endpoint, rather it is a tension to be explored and examined. Going against the grain of transparency, the gaining of authenticity for automated outputs can be a product of how unexplainable that system is. Advancing automation can itself be justified and legitimized by the very fact that its outputs cannot be tracked. The system justifies itself and its outputs through its own unknowability and through the presences of nonknowledge. This, of course, returns us again to Bataille's conceptual notion of nonknowledge. The movement beyond basic algorithmic structures and into systems like neural networks creates questions for the type of consciousness of nonknowledge. The problem here is how difficult things become when the concept of nonknowledge is applied to algorithmic thinking, especially when consciousness itself becomes part of the equation. It is not just about the already complex issue of grasping the consciousness or the absence of consciousness that Bataille spoke of, it is also about how the absence of consciousness is an aspect of the implementation of algorithmic consciousness itself. In accounting for the future of neural networks in a written statement from 1995, James

A. Anderson (1998: 262) claims that when it comes to neural networks 'the great question for the future, the most important question and the most difficult to solve is consciousness'. He wonders if this can only really be reached through vast and deep forms of complexity, adding that 'maybe there is just something special about systems of great complexity' (Anderson, 1998: 263). In this account the unknowable is a part of how automation might expand into consciousness. The pursuit of complexity required for this type of advancement inevitably creates tensions around what is known and what is knowable.

Anderson is not uncritical of what these reflections on consciousness might mean, particularly if a reductive version of consciousness were to become the model for this innovation. As he explains:

> The generation of consciousness through a nonlinear neural net that tries to solve the binding problem to provide more effective computation strikes me as unconvincing and almost insulting. Is that all consciousness is there for? ... Is there any special computational function that a conscious system can do that an unconscious one cannot? I suspect any particular well-defined function could always be produced by a mechanical system. (Anderson, 1998: 264)

So there is the sense that consciousness, however that might be defined, is an aim for such algorithmic thinking. At the same time, there is an open acknowledgement that consciousness is more than a simple set of processes, it is perceived to only exist within deep levels of complexity. In this way, consciousness itself is set up as a spot on the horizon of algorithmic thinking. This spot occurs where the complexity is great enough that consciousness of a type is achieved. It is perceived to be a spot that can only be reached through genuine complexity of reasoning and with the acceptance of the integration of nonknowledge. This type of vision is based not just on the known and the knowable but also on the active pursuit of the unknown and of actually, potentially, producing the unknowable.

There is one further comment in this past version of the future that stands out. Anderson (1998: 264) wonders if 'maybe computers are just as conscious as we are; they just don't talk about it'. This seemingly jovial and off-the-cuff provocation itself might indicate the presences of nonknowledge – being unaware of the level of the level of consciousness of computers – and it might also be suggestive of how nonknowledge is an established part of the framing of algorithmic thinking, a framing that is already translating into today's advancing forms of automation, machine learning and AI. There is a sense that something unknown needs to be made present in order for algorithmic thinking to advance. As the neural network expert Teuvo Kohonen (1998: 164) wonders, 'maybe a dream about a really cognitive machine is too far away, not least because it's difficult to define what these machines should be doing'.

Conclusion

In general terms, we might have the impression that algorithmic systems are based around rational and objective reasoning, or that they somehow are premised on a heightened transparency of decision-making, that is to say that we think of them as producing new knowns. Highlighting the types of tensions this creates, this chapter has explored the presences of nonknowledge in algorithmic thinking, taking this back through the development of neural networks. Algorithmic systems produce new unknowns and can even be built around the pursuit of increasing unexplainability. Expanding the learning of automated systems requires the making of mysteries. The presence of nonknowledge is required for some forms of algorithmic thinking to operate. There is, as is fitting with Bataille's outline, a desire for nonknowledge and its unstable limitlessness. The result is that in algorithmic thinking the knowable and the unknowable are in tension. These tensions are playing out in this set of relations in which algorithmic thinking produces both knowledge and nonknowledge at the same time. The questions about the increasing cognitive functioning of AI are then also questions about the increasing unexplainability of distributed systems and diffuse intelligence. One consequence to consider here is that, potentially, the more power these systems have and the more autonomous they are within social structures, the less explainable, trackable, comprehensible and ultimately responsible their decision-making may be.

Clearly, developing systems that are based upon nonknowledge and unexplainability creates vast ethical questions, particularly as such systems take on a role within the social world. Yet, this is not simply to say that algorithms can't be known. As Taina Bucher (2018) has already argued, algorithms are not necessarily black boxes that cannot be understood. Indeed, Bucher's (2018: 46) point is that even if they are, they still might be known through other means. And this, as she also contends, is if 'black box' is the right term anyway. Bucher (2018: 46) argues, that 'it is true that proprietary algorithms are hard to know, it does not make them unknowable'. Bucher (2018: 46) suggests that 'knowing' algorithms is to 'un-know' them. She explains that this requires us to 'engage more actively in addressing the things that seem to interfere with or keep us from knowing' (Bucher, 2018: 47). This is to analyse the mechanisms and means by which algorithms appear to be unknowable and to make this a focal point for examination. Part of the agenda that Bucher (2018: 64) is extending is to examine the presence of 'strategic unknowns'. This is where unknowability is not treated as a 'problematic'. The issues I am dealing with in this chapter reside at the limits of these strategic unknowns and where these strategic unknowns take the form of desired unexplainability. To refer back to Bataille, the levels of uneasiness with nonknowledge are far from uniform. When it comes to the nonknowledge in algorithmic systems, not all parties share a sense of disconcerting uneasiness with its presences. This variegated uneasiness with the presence of nonknowledge within neural networks is one means by which the tensions of algorithmic thinking might be elaborated, especially when nonknowledge is actively sought out and applied as part of the power of the system itself.

Clearly questions of the unknown properties of algorithms are crucial and they create the kinds of tensions that this book seeks to deal with. What Bucher (2018: 42) calls the 'problematic of the unknown', in which the way that algorithms are approached can itself be a barrier to their understanding, is a key part of the tensions of algorithmic thinking. Yet I have focused on a slightly different aspect of this within this particular chapter. I am arguing that the very aim of advancing algorithmic thinking is to create systems that are unexplainable and unknowable and that these systems are understood to require the presence of nonknowledge in order to function. Indeed, they are based around the desire for nonknowledge, to return to Bataille's terms, and the potentially unbounded and limitless possibilities that nonknowledge is seen to present. Automated systems of different types, and particularly those founded upon machine or deep learning and neural networks, do not eradicate or erode nonknowledge, rather they generate nonknowledges. They may create new knowns but in so doing they also cultivate unknowns.

Celia Lury (2021: 202) has recently argued that 'significant shifts in the epistemic infrastructure are underway' and that these can be understood as creating new 'problem spaces'. These problem spaces appear, Lury (2021: 203) claims, when generative changes are 're-distributive of cognitive agency'. The presence of nonknowledge within algorithmic thinking represents just such a problem space. It is the layering of depths within advancing algorithmic systems that represents one problem space, especially as these neural networks bring redistributions of agency and a growing sense of the unexplainability of decisions, outcomes and choices. This is not a fatalistic position, it is rather to think about the forms of nonknowledge that are being produced, how they function and the impact that they have. It is then to reflect on the presences of nonknowledge in algorithmic thinking and to try to grapple with the tensions that these presences create. What is not clear here is whether, to return once more to Bataille, this aim for unexplainability within the development of automation can be understood as a suspension of the will to know. In the next chapter I will argue that the will to know is instead supplanted or overridden by its powerful mutation as the *will to automate*.

6

Conclusion: Algorithmic Thinking and the Will to Automate

Algorithmic thinking is both elusive and everywhere. It is hard to see and impossible to avoid. While having significant power and influence, it is so intricately connected into social structures and everyday experiences that it goes almost unnoticed. Very little resides outside or untouched by algorithmic thinking and, crucially, its tensions. Acknowledging this hyped-up incorporation and unbending inescapability, Rosie DuBrin and Ashley E. Gorham (2021) have written of a form of 'algorithmic interpellation'. To engage with the social world is, frequently, to be interpellated into some form of algorithmic process. We are hailed by algorithmic systems; resisting the temptation to turn to face them is not really an option. Given this context, I'm not even sure that a complete or definitive understanding of contemporary algorithms and algorithmic thinking is possible. And anyway, any attempt to do so is likely to miss the dynamism and sheer variability of the issues and consequences. Alongside this, algorithmic thinking and algorithmic systems never seem to settle into place for long enough to be fully grasped. They are on the move, they change, they mutate and they are coded and recoded. The ideals of continual supposed advancement are built into them; they are codified in their very form. Even if algorithmic thinking were to freeze for long enough for a stable depiction to be sketched, the many different forms of algorithmic thinking would be too vast and contested to be fully mapped. As a result, any such picture is always likely to be incomplete and is always likely to miss the pulsing mobilities, momentums and dynamics of what automation and its various processes might incorporate or entail.

Shifting the focus to concentrate on this very mobility and dynamism is one way to proceed. In the case of this book, rather than seeing algorithmic processes as a fixed object of study I have instead taken some of the tensions of algorithmic thinking as my focus. This approach affords a perspective on the relational aspects of automation, the forces and frictions that define its forms and the ideals and future trajectories that are bound up within it.

Understanding these tensions may enable us to look through and beyond particular algorithmic developments and to create concepts that are of use outside isolated technological moments. Given that automation is really an ongoing process of *automating*, then a sense of the mobility and of the tensions that that mobility creates is vital.

With this in mind, this book is not intended as an endpoint. Instead, it is intended as only a refocusing. My hope is that it provides opportunities for exploring the unfolding relations of algorithmic systems and their implementations through this examination of the ongoing and unresolvable tensions. My hope is that some of the ideas in this book might be useful resources for understanding the variety of algorithmic modes of thinking as they happen and as they shift over time. The book is not intended to be exhaustive or to make a singular definitive statement on the tensions of algorithmic thinking. Instead it has explored two sets of competing and related tensions that are particularly prominent and that are unlikely to resolve. With these unfolding dynamics as a focus, I'd like to use this closing chapter to do two things. First I'll offer some reflections on the concepts developed across the previous chapters, exploring each in relation to the overall argument of the book. Then, second, I'll reflect on how we might understand the underpinning desires that are steering the direction and mobility of algorithmic thinking and its expansions. This closing section reflects on the *will to automate* to which these findings speak and which, I suggest, is a driving presence within the development and expansion of algorithmic thinking and its tensions.

Algorithmic tensions

Overall, to reiterate, I have argued that algorithmic thinking should be understood through its tensions. More than this though, it can only really be understood as being tense. Algorithmic thinking is not fixed or stable and nor is it informed by a single set of shared ideals, unified principles or a single logic. I don't imagine that anyone was particularly expecting to find that algorithmic thinking is uniform or monological, however established algorithmic systems may seem. Despite the dynamism, the forces shaping the direction of algorithmic thinking and the tensions this creates can potentially be captured and explored. As I have already suggested, the dynamism itself can then become the object of study as can the tensions that define that dynamism. There are likely to be many such tensions, especially as different types of automation pull and push at conventions, norms and structures. In this book, as I explained, among the fraughtness I have attempted to explore two particularly strong tensions of algorithmic thinking. On the one side, I looked at the tensions across the spectrum of the human and humanlessness. On the other side, I looked at the tensions between the known and the unknown. These represent a starting point from which to understand the specificities of the different types of automation that are unfolding. There will be other tensions. Many other tensions. And beyond that, even the two sets of tensions I have focused upon in this book themselves have many dimensions and facets within them. It was with this in mind that I began to try to carve out four concepts that might allow these some aspects of these tensions and the forces provoking them to be further explored. The key concepts I have outlined in this book are not exhaustive of what is happening, rather they can be used to focus attention on certain aspects of these dynamics and to explore certain spaces within the vectors of these algorithmic tensions. These four concepts can be worked with in isolation or, to see the interconnection of the tensions, can be used in combinations.

First, the concept of *posthuman security* (see Chapter 2) emerged from an analysis of the envisioned potentials and application blockchain. As a concept, posthuman security is intended as a focal point for thinking about the reasons and means by which a greater level of humanlessness is prioritized and sought out within processes of automation. Posthuman security is intended to provide a focal point for thinking about how notions of vulnerability and riskiness are a part of how the human is downgraded within the meshing of different agencies. The pursuit of posthuman security is an attempt to erase the human in order to remove those perceived vulnerabilities and to achieve trust in these systems. This is founded in the technological dream of a secure society. It can be explored in its own right, but the key point is that posthuman securities are always likely to be in tension with those pushing against them, especially where there is an impulse to protect the qualities that are deemed to be more human. Which leads us to a further concept ...

Second, and in response to the push of humanlessness, it is on the point of human retention that the concept of *overstepping* was intended to make an intervention (see Chapter 3). This concept highlights attempts to actively position the human within algorithmic systems and to circumscribe agency within those systems. The perceived limits of automation, by which I mean the various lines that are drawn around what is deemed to be an acceptable of level of algorithmic intervention, are not fixed or stable. Instead these perceived limits are being actively managed and redrawn. The concept of overstepping is intended to focus attention on how these limits are managed and on how an understanding of the limits might itself be used to breach or stretch those limits. Part of understanding the mobility and dynamism of automation is to understand its boundaries and how these perimeters come to be stepped over. The analysis of overstepping provided in Chapter 3 explored in particular how quite subtle and embedded notions of what might constitute *too much automation* can be a part of how automation then becomes established. A sense of what would be regarded as too much automation can even be a part of how automated processes and systems expand. The avoidance of overstepping allows this expansion to occur. In short, algorithmic thinking can incorporate a sense of its own limits and maintain an eye on how they might be managed and pushed back.

In the second half of the book the attention moved away from the focus on the tensions between the human and humanlessness and focused instead on those that arise between the known and the unknown. Two concepts were developed to try to deal with aspects of this tension. Drawing on Katherine Hayles' concept of the cognizer, Chapter 4 looked at how *dreams of super cognizers* were deployed to expand what is known and what is knowable. Indeed, the concept of the super cognizer is intended to provide angles for thinking about what the possibilities are and how an engagement with making things knowable is a fundamental part of algorithmic thinking. As an extension of the cognizer, the super cognizer draws the analysis towards what can be known and how algorithmic systems alter perception to stretch what is known or what is knowable. The super cognizer acts as a bridge or gateway for thinking about how current forms of automation are actively developed and how the ideals and logics of algorithmic thinking become embodied in particular technologies. The dreams of the super cognizer are based in seeing the known as something that is sought after and as something in which ideas about what is desirable knowledge are being framed. This concept is intended to provide perspectives on how algorithmic thinking is tied up with notions of what can be known and what is knowable. The super cognizer, as Chapter 4 also explored, might be seen to be active in the ongoing reconfiguration of the relations between the human and the machine. Super cognizers, as we saw, are implicated in the expansion of the field of vision of automated analytics, opening up what it is that can be analysed and how that knowing is turned into logistics, decisions and actions. The politics of the human and the politics of knowing are not necessarily separable, as the discussions of the super cognizer indicated.

On the other side of these forces, aimed at understanding the tensions that occur around the unknowable, the final concept dealt with was *nonknowledge*. Drawing upon and adapting the writings of Georges Bataille, the concept of nonknowledge is intended to explore how the limits of knowledge and the known are established and the role that not-knowing can play in the functioning of social life, relations and notions of authenticity. The concept of nonknowledge was developed in Chapter 5 with the aim of thinking about how processes of automation create new forms of nonknowledge. With neural networks taken as a focal point for thinking about how the capabilities of automation are accompanied by new forms of unexplainability, the chapter examined the desire for nonknowledge that has long been established in such innovations. Indeed, the chapter explored how nonknowledge was established form the early stages of neural networks and deep learning. The use of the concept of nonknowledge is intended to sensitize the analysis to the very pursuit of unexplainability that is embedded in more advanced forms of automation – as embodied in the depth of neural networks and the hidden layers they contain. Chapter 5 looked at how algorithmic thinking operates through the presences of nonknowledge. In the case of neural networks there is an active pursuit of nonknowledge that creates tensions between the knowable and the unknowable.

The extension of the known was explored through the concept of the super cognizer, whereas the notion of the presences of nonknowledge has been used to reflect on how knowing is placed in tension with the extension of the unknown. Advances in algorithmic thinking, I argued, are premised and defined by the presence of nonknowledge. Nonknowledge is one way into understanding the blind spots that populate algorithmic thinking and how those blind spots can be actively engineered into these systems. The presences of nonknowledge in automation are focused in this particular case on how algorithmic thinking may seek to know while also producing other types of unknown. It can even be the case that such systems draw their authenticity from nonknowledge.

In this book I have treated algorithmic thinking as both the thinking is done *about* algorithms and the apparent thinking done *by them*. Returning again to Henri Lefebvre's points about modernity, with which I opened this book (see Chapter 1), we might wonder what type of 'new life' is encapsulated within algorithmic thinking. The tensions explored in this book push and pull, shaping the direction that unfolding forms of automation take. The prominent myths of the new life are no doubt algorithmic in their forms and ideals, and this presents us with questions about how those tensions implicate both the present and the future. Algorithmic thinking is well established in shaping social life and so, therefore, are its tensions. My argument is that algorithmic thinking can only really be understood through an analysis of the tensions that define it. These tensions afford the mobility and dynamism of algorithmic thinking; they also shape its direction and expansions. Algorithmic thinking is tense; it never escapes its tensions, rather it is sculpted by their very presence. Algorithmic thinking is in an almost constant state of motion. It may keep shifting and mutating, but the tensions that define it are likely to be a relative constant. It is here that we can come to appreciate the myths of new life that define the direction and implementation of algorithmic thinking. It is also where we can see the forces that shape the forms that algorithmic thinking might take.

The will to automate

Let me conclude with one final and more general observation. Exploring the tensions of algorithmic thinking also has the potential to tell us something about its underlying principles, ideals and drives. We could go as far as to conclude that there is a palpable *will to automate* behind the establishment, spread and dynamics of algorithmic thinking. Founded in a drive to expand its presence, the will to automate pushes algorithmic thinking outwards, making it more intense and also stretching its reach. It is a mode of thinking that is based upon identifying ever more practices, means and processes that might be rendered open to automation. The underpinning objective is to automate more and more aspects of the social world. The question then is how to understand the desire to automate that gives algorithmic thinking its dynamic properties. Its relation to the will to know provides a starting point.

In the 1970–71 course summary that followed his lecture series from that academic year, Michel Foucault (2013: 225) indicated that the study of the will to know requires a 'distinction between knowledge-*savoir* and knowledge-*connaissance*'. This, he explains, is based on the difference 'between will to know (savoir) and will to truth'. More specifically, Foucault is concerned with how subjects are positioned in 'relation to that will' (Foucault, 2013: 226). Analysing the practices and discourses that make the will to know possible, he argued, is a crucial part of understanding it. Foucault explains that his concern with the will to know is focused on understanding 'that which makes possible its existence' (Foucault, 2013: 6). In other words, he is interested in how that will to know comes about and how it comes to take that form. The conditions of possibility are then bound up with knowledge itself. The same may well be said of automation, with the conditions of possibility bound up with its form.

Making a direct gesture towards the particularly entangled nature of this will to know, Foucault observes that 'in its nature, action, and power, the desire to know is not outside the knowledge it desires' (Foucault, 2013: 16). The very will or desire to know is part of this knowing – knowing of that desire is bound up in the will to know. This is not dissimilar to the type of reflective aspects of algorithmic thinking that have crept up across this book's chapters, especially where the desire to know is based upon expanding algorithmic thinking further. Foucault (2013: 16) goes on to emphasize that 'the desire to know is no more than a game of knowledge in relation to itself, it does no more than show its genesis, delay, and movement; desire is knowledge deferred, but made visible in the impatience of the suspense in which it is held' (Foucault, 2013: 16). There is a desire to gain knowledge that is deferred, to impatiently move beyond what is known.

The immanence of the desire for knowledge is something Foucault addresses further, highlighting how it has a defining presence with the will to know. The will to know feeds off itself. 'The desire to know', Foucault (2013: 6) explains, 'is in its nature already something like knowledge, something belonging to knowledge'. Echoing some of the discussions of nonknowledge in Chapter 5 and overstepping in Chapter 3, the limits of knowing are themselves a required part of the desire to know more. The argument here, as Foucault (2013: 16–17) puts it, is that 'knowledge is at once its object, its end, and its material'. The desire to know is then, in this particular account from Foucault, bound up with the knowledge it creates and the way its own presence is understood, it is inherent within it. The will to know requires knowledge in order for that desire to be expressed. More than this though, the desire to know is also a type of knowledge itself. The result of this set of relations is that 'desire is no longer cause, but knowledge that becomes cause of itself' (Foucault, 2013: 18). The will to know drives itself on. It creates knowledge which then creates the desire for further knowledge, and so on. The question this then poses in the context of this book is how the establishment of automation might then drive the desire for more automation.

Foucault (2013: 22) places knowledge in the foundations of the desire to know, arguing that 'there was knowledge at the root of the desire, even before it manifests itself and starts to function ... a knowledge already there on the basis of which the desire could function'. In short, it would seem, the desire driving the will to know does not arrive from nowhere; it is built out of the knowledge that this drive creates. This is suggestive of not just how the will to know can drive what is known, but also how what is known can drive the desire to know and the direction in which the conditions of knowing might develop. This can then mean that the will to know can, in Foucault's view, be an embedded institutional presence. Reflecting on this aspect of the agenda he is setting out, Foucault (1971: 11) adds that 'this will to knowledge, thus reliant upon institutional support and distribution, tends to exercise a sort of pressure, a power of constraint upon other forms of discourse'. And so, supported and informed by institutional structures, the will to know has a powerful presence in defining the constraints and possibilities of knowledge. It also creates limits and constraints on alternative reasoning. We may reflect then on how a will to know can constrain and make possible certain types of automation; we may also wonder how the conditions of possibility and embedded knowledge may shape the underpinning desire and direction of algorithmic thinking.

One conclusion that we could draw from this is that with the drive to expand, integrate and advance automation, we may not only be looking at a certain form of the will to know but we may also, within this wider concept, be seeing a *will to automate*. The *will to know* could well be provoking and guiding the *will to automate* (Chapters 4 and 5 would both suggest that this is the case). Yet, the will to automate is a particular and distinct incarnation of the will to know – a mutation of it. We even saw in Chapter 5's discussions of nonknowledge how the will to know might be usurped by a will to automate, with a desire to know being sidelined by the stronger desire to automate without knowing. The will to automate is the version of the will to know that occurs where an algorithmic new life is conjured, envisioned and implemented. Those visions fuel the desire to automate further.

What we might take from this is that as the will to know shifts so too might the forms and direction that the will to automate takes. If we reflect back across the chapters of this book, the version of the will to automate that is implicit within its pages has some broad yet identifiable features: it is expansionary and seeks to increase the range and depth of automation where possible; it adapts its practices and knowledge to achieve expansion and intensification; it works through a logic of calculation, judgement and the framing of algorithms as providing advancements to decision-making; it is observational, surveillant and self-organizing (within the limits it seeks to establish and to breach); it seeks authority and legitimacy through combinations of devices along with perceived speediness and efficiency; it is founded in a regime of promises that are invested in envisioning the next steps and finding ways to establish them within current devices and systems; and, finally, its desires are often geared towards the potential of new knowledges and the possibility of the limitless unknown. I could continue, but these are some of the most observable features of the will to automate based upon the analysis I have provided in the previous chapters. Like the will to know, studying the will to automate also requires an analysis of the conditions that make it possible, including institutional, commercial and organizational structures and discourses. In part at least, this is what I have sought to do in this book.

Of course, it will be important to explore the will to automate as historically contingent and changeable. This book gives glimpses of this, but, because of its aims, these are only snapshots. The very changes in the will to automate as well as the transformations to automation itself should be explored together. Further to this, it may well require an exploration of how the will to know takes certain forms and becomes embodied in certain ways within types of automated systems. One question I am leaving open here is how the will to automate changes over time, what forms and varieties it might take and the types of power that might operate to shape it. More than this though, it is a question of the conditions, practices and discourses that have made these forms of the will to automate possible. These are especially important as we consider how the will to automate feeds back into its own desires. These are the conditions of possibility in which algorithmic thinking operates. Knowledge, Foucault (2013: 210) points out, 'rests on a network of relations'. The question then is how to grasp that network of relations in the case of developing algorithmic systems and structures. The logic of algorithmic thinking and its deployment in particular forms of automation might then itself drive and shape the will to automate. And so, in this sense at least, this book is part of a bigger project that might seek to comprehend the will to automate as both a form of knowledge and a set of conditions that afford and direct that knowledge. An understanding of the tensions of algorithmic thinking requires an analysis of the *will to automate* that drives it. Such a project would essentially seek to understand and contextualize algorithmic thinking in terms of the particularities of its desires, impulses and the manifestation of its conditions of possibility.

References

Aerum (2019) 'Empowering tokenization and programmable finance', Aerum, https://aerum.com (accessed 6 April 2019).

Amoore, L. (2013) *Politics of Possibility: Risk and Security Beyond Probability*, Durham, NC: Duke University Press.

Amoore, L. (2019) 'Doubt and the algorithm: on the partial accounts of machine learning', *Theory, Culture & Society*, 36(6): 147–69.

Amoore, L. (2020) *Cloud Ethics: Algorithms and the Attributes of Ourselves and Others*, Durham, NC: Duke University Press.

Amoore, L. and Piotukh, V. (eds) (2016) *Algorithmic Life: Calculative Devices in the Age of Big Data*, London: Routledge.

Amoore, L. and Piotukh, V. (2019) 'Interview with N. Katherine Hayles', *Theory, Culture & Society*, 36(2): 145–55.

Anderson, J.A. (1998) 'James A. Anderson', in J.A. Anderson and E. Rosenfeld (eds) *Talking Nets: An Oral History of Neural Networks*, Cambridge, MA: MIT Press, pp 239–64.

Anderson, J.A. and Rosenfeld, E. (eds) (1998) *Talking Nets: An Oral History of Neural Networks*, Cambridge, MA: MIT Press.

Anderson, N.A. and Stenner, P. (2020) 'Social immune mechanisms: Luhmann and potentialization technologies', *Theory, Culture & Society*, 37(2): 79–103.

Andrejevic, M. (2013) *Infoglut: How Too Much Information Is Changing the Way We Think and Know*, London: Routledge.

Andrejevic, M. (2020) *Automated Media*, London: Routledge.

Art Market Guru (2018) 'ARTCHAIN: Interview with Kay Sprague and Cameron MacQueen', Art Market Guru, www.artmarket.guru/le-journal/interviews/artchain-kay-sprague/ (accessed 27 October 2020).

Art Market Guru (2019) 'Blockchain companies in the art market', Art Market Guru, www.artmarket.guru/le-journal/blockchain/blockchain-companies/ (accessed 6 April 2019).

Asp, K. (2019) 'Autonomy of artificial intelligence, ecology, and existential risk', in T. Heffernan (ed) *Cyborg Futures: Cross-Disciplinary Perspectives on Artificial Intelligence and Robotics*, London: Palgrave Macmillan, pp 63–88.

Atkinson, R. and Blandy, S. (2016) *Domestic Fortress: Fear and the New Home Front*, Manchester: Manchester University Press.

Bailey, J. (2018) 'Why use blockchain provenance for art?', *Artnome*, 29 January 2018, www.artnome.com/news/2018/1/26/why-use-blockch ain-provenance-for-art (accessed 22 April 2021).

Basar, S. (2020) 'BlackRock to launch Aladdin Climate', *Markets Media*, 13 October 2020, www.marketsmedia.com/blackrock-to-launch-aladdin-clim ate/ (accessed 17 November 2020).

Bataille, G. (2001) *The Unfinished Systems of Nonknowledge*, edited with an introduction by S. Kendall, translated by M. Kendall and S. Kendall, Minneapolis, MN: University of Minnesota Press.

Bechmann, A. and Bowker, G.C. (2019) 'Unsupervised by any other name: hidden layers of knowledge production in artificial intelligence on social media', *Big Data & Society*, 6(1): 1–11.

Beer, D. (2017) 'The social power of algorithms', *Information, Communication & Society*, 20(1): 1–13.

Beer, D. (2019a) *The Data Gaze: Capitalism, Power and Perception*, London: Sage.

Beer, D. (2019b) *Georg Simmel's Concluding Thoughts: Worlds, Lives, Fragments*, London: Palgrave Macmillan.

Beer, D. (2020) 'Trapped in a code: the fight over our algorithmic future', *Open Democracy*, 21 August 2020, www.opendemocracy.net/en/open democracyuk/trapped-in-a-code-the-fight-over-our-algorithmic-future/ (accessed 30 April 2021).

Beer, D. (2021) 'Archive fever revisited: algorithmic archons and the ordering of social media', in L.A. Lievrouw and B.D. Loader (eds) *Routledge Handbook of Digital Media and Communication*, London: Routledge, pp 99–112.

Beer, D., Redden, J., Williamson, B. and Yuill, S. (2019) 'Landscape summary: online targeting – what is online targeting, what impact does it have, and how can we maximise the benefits and minimise harms?', *Centre for Data Ethics and Innovation*, https://assets.publishing.service.gov.uk/gov ernment/uploads/system/uploads/attachment_data/file/819057/Landsca pe_Summary_-_Online_Targeting.pdf (accessed 3 March 2020).

Binder, W. (2021) 'AlphaGo's deep play: technological breakthrough as social drama', in J. Roberge and M. Castelle (eds) *The Cultural Life of Machine Learning: An Incursion into Critical AI Studies*, London: Palgrave Macmillan, pp 167–95.

Birchall, C. (2011) 'Introduction to "secrecy and transparency": the politics of opacity and openness', *Theory, Culture & Society*, 28(7–8): 7–25.

Bjerg, O. (2016) 'How is Bitcoin money?', *Theory, Culture & Society*, 33(1): 53–72.

BlackRock (2020a) 'Aladdin: risk managers', BlackRock, www.blackrock. com/aladdin/benefits/risk-managers (accessed 17 August 2020).

BlackRock (2020b) 'Aladdin', BlackRock, www.blackrock.com/aladdin# (accessed 17 August 2020).

BlackRock (2020c) 'Aladdin enterprise', BlackRock, www.blackrock.com/aladdin/offerings/aladdin-overview (accessed 17 August 2020).

BlackRock (2020d) 'Aladdin wealth', BlackRock, www.blackrock.com/aladdin/offerings/aladdin-wealth (accessed 17 August 2020).

BlackRock (2020e) 'Aladdin organizations', BlackRock, www.blackrock.com/aladdin/benefits/organizations (accessed 17 August 2020).

BlackRock (2020f) 'BlackRock institutions: Aladdin', BlackRock, www.blackrock.com/institutions/en-zz/solutions/aladdin (accessed 16 November 2020).

Borch, C. (2016) 'High-frequency trading, algorithmic finance and the Flash Crash: reflections on eventalization', *Economy & Society*, 45(3–4): 350–78.

Boyd, D. and Crawford, K. (2012) 'Critical questions for big data: provocations for a culture, technological, and scholarly phenomenon', *Information, Communication & Society*, 15(5): 662–79.

Bradley, D. (2019) 'Locking down your smart home with blockchain', *TechXplore*, 12 December 2019, https://techxplore.com/news/2019-12-smart-home-blockchain.html (accessed 21 October 2020).

Braidotti, R. (2013) *The Posthuman*, Cambridge: Polity.

Braidotti, R. (2019) *Posthuman Knowledge*, Cambridge: Polity.

Bromand, D., Gustafsson, D., Mitic, R. and Mennicken, S. (2020) 'Systems and methods for enhancing responsiveness to utterances having detectable emotion', Spotify AB, US 2020/0219511, https://patentimages.storage.googleapis.com/81/08/30/5be9877d1791d5/US20200219511A1.pdf (accessed 16 February 2021).

Bucher. T. (2017) 'The algorithmic imaginary: exploring the ordinary affects of Facebook algorithms', *Information, Communication & Society*, 20(1): 30–44.

Bucher, T. (2018) *If ... Then: Algorithmic Power and Politics*, Oxford: Oxford University Press.

Campbell-Verduyn, M. (2018) 'Introduction: what are blockchains and how are they relevant to governance in the global political economy', in M. Campbell-Verduyn (ed) *Bitcoin and Beyond: Cryptocurrencies, Blockchains, and Global Governance*, London: Routledge, pp 1–24.

Cave, S. and Dihal, K. (2019) 'Hopes and fears for intelligence machines in fiction and reality', *Nature Machine Intelligence*, 1: 74–8.

Chapman, S. (2021) 'The new digital gold: why the recent Bitcoin crash won't halt the growth in crypto assets', *New Statesman*, 9–15 July 2021, pp 37–41.

Chow-White, P., Lusoli, A., Phan, V.T.A. and Green, S.E. (2020) '"Blockchain good, Bitcoin bad": the social construction of blockchain in mainstream and specialized media', *Journal of Digital Social Research*, 2(2): 1–27.

Chun, W.H.K. (2021) *Discriminating Data: Correlation, Neighborhoods, and the New Politics of Recognition*, Cambridge, MA: MIT Press.

Collins, H. (2018) *Artificial Intelligence: Against Humanity's Surrender to Computers*, Cambridge: Polity Press.

Cooper, L.N. (1998) 'Leon N. Cooper', in J.A. Anderson and E. Rosenfeld (eds) *Talking Nets: An Oral History of Neural Networks*, Cambridge, MA: MIT Press, pp 71–95.

Cowan, J.D. (1998) 'Jack D. Cowan', in J.A. Anderson and E. Rosenfeld (eds) *Talking Nets: An Oral History of Neural Networks*, Cambridge, MA: MIT Press, pp 97–126.

Crawford, K. (2015) 'Can and algorithm be agonistic? Ten scenes from life in calculated publics', *Science, Technology, & Human Values*, 41(1): 77–92.

Crawford, K. (2021) *Atlas of AI: Power, Politics, and the Planetary Costs of Artificial Intelligence*, New Haven, CT and London: Yale University Press.

Curtis, A. (2014) 'Now Then', *BBC*, 25 July 2014, www.bbc.co.uk/blogs/adamcurtis/entries/78691781-c9b7-30a0-9a0a-3ff76e8bfe58 (accessed 25 February 2021).

Davies, W. (2016) *The Happiness Industry*, London: Verso.

De Filippi, P. and Hassan, S. (2016) 'Blockchain technology as a regulatory technology: from code is law to law is code', *First Monday*, 21(12), https://doi.org/10.5210/fm.v21i12.7113 (accessed 22 February 2021).

Delfanti, A. and Frey, B. (2020) 'Humanly extended automation of the future of work seen through Amazon patents', *Science, Technology & Human Values*, online first, https://doi.org/10.1177/016224392 0943665 (accessed 22 February 2021).

DeNora, T. (2000) *Music in Everyday Life*, Cambridge: Cambridge University Press.

Dodd, N. (2018) 'The social life of Bitcoin', *Theory, Culture & Society*, 35(3): 35–56.

Dodge, M. and Kitchin, R. (2009) 'Software, objects, and home space', *Environment and Planning A*, 41(6): 1344–65.

DuBrin, R. and Gorham, A.E. (2021) 'Algorithmic interpellation', *Constellations*, 28(2): 176–91.

Dunn, W. (2018) 'Meet Aladdin, the computer "More powerful than traditional politics"', *New Statesman*, 6 April 2018, www.newstatesman.com/spotlight/2018/04/meet-aladdin-computer-more-powerful-traditio nal-politics (accessed 16 November 2020).

DuPont, Q. (2019) *Cryptocurrencies and Blockchains*, Cambridge: Polity.

Dyer-Witheford, N., Kjøsen, A.M. and Steinhoff, J. (2019) *Inhuman Power: Artificial Intelligence and the Future of Capitalism*, London: Pluto Press.

Elhanani, Z. (2018) 'How blockchain changed the art world in 2018', *Forbes*, 17 December 2018, www.forbes.com/sites/zoharelhanani/2018/12/17/how-blockchain-changed-the-art-world-in-2018/ (accessed 8 July 2019).

Elliot, A. (2019) *The Culture of AI: Everyday Life and the Digital Revolution*, London: Routledge

Faria, I. (2019) 'Trust, reputation and ambiguous freedoms: financial institutions and subversive libertarians navigating blockchain, markets, and regulation', *Journal of Cultural Economy*, 12(2): 119–32.

Faustino, S. (2019) 'How metaphors matter: an ethnography of blockchain-based re-descriptions of the world', *Journal of Cultural Economy*, 12(6): 478–90.

Fazi, B. (2021) 'Beyond human: deep learning, explainability and representation', *Theory, Culture & Society*, 38(7–8): 55–77.

Ferrari, F. and Graham, M. (2021) 'Fissures in algorithmic power: platforms, code, and contestation', *Cultural Studies*, 35(4–5): 814–32.

Flyverbom, M. (2019) *The Digital Prism: Transparency and Managed Visibilities in a Datafied World*, Cambridge: Cambridge University Press.

Foucault, M. (1971) 'Orders of discourse', *Social Science Information*, 10(2): 7–30.

Foucault, M. (1980) *Power/Knowledge: Selected Interview and Other Writings, 1972–1977*, edited by C. Gordon, translated by C. Gordon, J. Mepham, K. Soper and L. Marchall, London: Pantheon Books.

Foucault, M. (2013) *Lectures on the Will to Know: Lectures at the Collège de France 1970–1971*, Basingstoke: Palgrave Macmillan.

Freeman, W.J. (1998) 'Walter J. Freeman', in J.A. Anderson and E. Rosenfeld (eds) *Talking Nets: An Oral History of Neural Networks*, Cambridge, MA: MIT Press, pp 23–42.

Gagne, Y. (2018) 'How the blockchain is changing the art market', *Fast Company*, 10 December 2018, www.fastcompany.com/90265029/how-the-blockchain-is-changing-the-art-market (accessed 6 April 2019).

Gane, N. (2005) 'Radical post-humanism: Friedrich Kittler and the primacy of technology', *Theory, Culture & Society*, 22(3): 25–41.

Grossberg, S. (1998) 'Stephen Grossberg', in J.A. Anderson and E. Rosenfeld (eds) *Talking Nets: An Oral History of Neural Networks*, Cambridge, MA: MIT Press, pp 167–96.

Hall, A. (2017) 'Decisions at the data border: discretion, discernment and security', *Security Dialogue*, 48(6): 488–504.

Hardesty, L. (2017) 'Explained: neural networks', *MIT News*, 14 April 2017, https://news.mit.edu/2017/explained-neural-networks-deep-learning-0414 (accessed 4 December 2020).

Hayes, A. (2019) 'The socio-technological lives of Bitcoin', *Theory, Culture & Society*, 36(4): 49–72.

Hayles, N.K. (1999) *How We Became Posthuman*, Chicago, IL: University of Chicago Press.

Hayles, N.K. (2005) *My Mother Was a Computer: Digital Subjects and Literary Texts*, Chicago, IL: Chicago University Press.

Hayles, N.K. (2006) 'Unfinished work: from cyborg to cognisphere', *Theory, Culture & Society*, 23(7–8): 159–66.

Hayles, N.K. (2017) *Unthought: The Power of the Cognitive Nonconscious*, Chicago, IL: Chicago University Press.

Hecht-Nielsen, R. (1998) 'Robert Hecht-Nielsen', in J.A. Anderson and E. Rosenfeld (eds) *Talking Nets: An Oral History of Neural Networks*, Cambridge, MA: MIT Press, pp 293–313.

Heffernan, T. (2019) '"Fiction meets science": ex machina, artificial intelligence, and the robotics industry', in T. Heffernan (ed) *Cyborg Futures: Cross-Disciplinary Perspectives on Artificial Intelligence and Robotics*, London: Palgrave Macmillan, pp 127–40.

Henderson, R. and Walker, O. (2020) 'FT big read: Blackrock', *Financial Times*, 25 February 2020, p 11.

Herian, R. (2018) 'Taking blockchain seriously', *Law and Critique*, 29(2): 163–71.

Hinton, G.E. (1998) 'Geoffrey E. Hinton', in J.A. Anderson and E. Rosenfeld (eds) *Talking Nets: An Oral History of Neural Networks*, Cambridge, MA: MIT Press, pp 361–84.

IBM (2020) 'Natural language processing (NLP)', *IBM*, 2 July 2020, www.ibm.com/cloud/learn/natural-language-processing (accessed 3 March 2022).

IoTex (2018) 'Blockchain for smart homes', *IoTex Official Channel*, 13 October 2018, www.youtube.com/watch?v=11PmTCAMtRQ (accessed 19 October 2020).

Jacobsen, B. (2021) 'Regimes of recognition on algorithmic media', *New Media & Society*, online first , https://doi.org/10.1177/14614448211053555 (accessed 30 August 2022).

Jacobsen, B. and Beer, D. (2021) *Social Media and the Automatic Production of Memory: Classification, Ranking and the Sorting of the Past*, Bristol: Bristol University Press.

Jones, M. (2018) 'Walmart goes after patent for smart appliance management on the blockchain', *The Block*, 3 August 2018, https://blockchaintechnol ogy-news.com/2018/08/walmart-goes-after-patent-for-smart-appliance-management-on-the-blockchain/ (accessed 19 October 2020).

Keim, R. (2020) 'How many hidden layers and hidden nodes does a neural network need', *All About Circuits*, 31 January 2020, www.allaboutcircuits.com/technical-articles/how-many-hidden-layers-and-hidden-nodes-does-a-neural-network-need/ (accessed 3 March 2022).

Kitchin, R. (2014) *The Data Revolution: Big data, Open Data, Data Infrastructures & Their Consequences*, London: Sage.

Kohonen, T. (1998) 'Teuvo Kohonen', in J.A. Anderson and E. Rosenfeld (eds) *Talking Nets: An Oral History of Neural Networks*, Cambridge, MA: MIT Press, pp 145–66.

Koopman, C. (2019) *How We Became Our Data: A Genealogy of the Informational Person*, Chicago, IL: Chicago University Press.

Lashuk, A. (2020) 'Blockchain knocks at your door: what solutions can DLT tools bring to smart homes?', *Openledger Insights*, 23 January 2020, https://openledger.info/insights/blockchain-in-smart-homes/ (accessed 19 October 2020).

Latimer, J. and López Gómez, D. (2019) 'Intimate entanglements: affects, more-than-human intimacies and the politics of relations in science and technology', *The Sociological Review*, 67(2): 247–63.

Lefebvre, H. (1995) *Introduction to Modernity*, London: Verso.

Lepage-Richer, T. (2021) 'Adversariality in machine learning systems: on neural networks and the limits of knowledge', in J. Roberge and M. Castelle (eds) *The Cultural Life of Machine Learning: An Incursion into Critical AI Studies*, London: Palgrave Macmillan, pp 197–225.

Lim, K. (2016) 'Blocks, chains and hashes', *Business Times*, 11 May 2016, www.businesstimes.com.sg/sites/default/files/attachment/2016/05/11/BT_20160511_KEN11_2272708.pdf (accessed 22 February 2021).

Long Jr, J.H. (2019) 'Evolution ain't engineering: animals, robots, and the messy struggle for existence', in T. Heffernan (ed) *Cyborg Futures: Cross-Disciplinary Perspectives on Artificial Intelligence and Robotics*, London: Palgrave Macmillan, pp 17–34.

Lupton, D. (2015) *Digital Sociology*, London: Routledge.

Lupton, D. (2020) *Data Selves: More-Than-Human Perspectives*, Cambridge: Polity Press.

Lury, C. (2021) *Problem Spaces: How and Why Methodology Matters*, Cambridge: Polity.

Lustig, C. and Nardi, B. (2015) 'Algorithmic authority: the case of Bitcoin', 48th Hawaii International Conference of Systems Sciences, https://doi.org/10.1109/HICSS.2015.95 (accessed 21 April 2021).

MacDonald-Korth, D., Lehdonvirta, V. and Meyer, E.T. (2018) *'Art Market 2.0: Blockchain and Financialisation in Visual Arts'*, The Alan Turing Institute: London, www.oii.ox.ac.uk/publications/blockchain-arts.pdf (accessed 6 April 2019).

Mackenzie, A. (2017) *Machine Learners: Archaeology of a Data Practice*, Cambridge, MA: The MIT Press.

MacKenzie, D. (2019) 'Pick a nonce and try a hash: Donald MacKenzie on Bitcoin', *London Review of Books*, 18 April 2019, pp 35–8.

Mager, A. and Katzenbach, C. (2021) 'Future imaginaries in the making and governing of digital technology: multiple, contested, commodified', *New Media & Society*, 23(2): 223–36.

Markham, A. (2021) 'The limits of the imaginary: challenges to intervening in future speculations of memory, data, and algorithms', *New Media & Society*, 23(2): 382–405.

Marshall, B. (2018) 'How are transactions validated?', *Medium*, 2 February 2018, https://medium.com/@blairlmarshall/how-do-miners-validate-transactions-c01b05f36231 (accessed 22 April 2021).

Matzner, T. (2019) 'The human is dead – long live the algorithm! Human-algorithmic ensembles and liberal subjectivities', *Theory, Culture & Society*, 36(2): 123–44.

Mauricio (2017) 'What you may not know: the software that is more powerful than Windows or even the US Government', *Medium*, 24 August 2017, https://medium.com/twogap/what-you-may-not-know-the-softw are-that-is-more-powerful-than-windows-or-even-us-goverment-8f2e0 8822673 (accessed 25 February 2021).

McClintock, M. (2017) 'Blockchain workflow automation: why you should embrace it', *ProcessMaker*, 20 December 2017, www.processmaker.com/ blog/blockchain-workflow-automation-why-you-should-embrace-it/ (accessed 5 November 2020).

McStay, A. (2016) 'Empathic media and advertising: industry, policy, legal and citizen perspectives (the case for intimacy)', *Big Data & Society*, 3(2): 1–11.

McStay, A. (2018) *Emotional AI: The Rise of Empathic Media*, London: Sage.

McStay, A. and Rosner, G. (2021) 'Emotional artificial intelligence in children's toys and devices: ethics, governance and practical remedies', *Big Data & Society*, 8(1): 1–16.

Mead, C. (1998) 'Carver Mead', in J.A. Anderson and E. Rosenfeld (eds) *Talking Nets: An Oral History of Neural Networks*, Cambridge, MA: MIT Press, pp 127–44.

Mendon-Plasek, A. (2021) 'Mechanized significance and machine learning: why it became thinkable and preferable to teach machines to judge the world', in J. Roberge and M. Castelle (eds) *The Cultural Life of Machine Learning: An Incursion into Critical AI Studies*, London: Palgrave Macmillan, pp 31–78.

Mire, S. (2018) '12 startups using blockchain to transform the art industry [Market Map]', *Disruptor*, 25 December 2018, www.disruptordaily.com/ blockchain-market-map-art/ (accessed 9 November 2020).

Mitchell, W.J. (2005) *Placing Words*, Cambridge, MA: MIT Press.

Murphy, H. (2018) 'Stablecoins spring up as crypto tries to shed Wild West image', *Financial Times*, 17–18 November 2018.

Naughton, J. (2018) 'Don't let bitcoin greed blind you to the potential of blockchain technology', *The Observer*, 13 May 2018, p 27.

Nayar, P.K. (2014) *Posthumanism*, Cambridge: Polity Press.

Nelms, T.C., Maurer, B., Swartz, L. and Mainwaring, S. (2018) 'Social payments: innovation, trust, bitcoin, and the sharing economy', *Theory, Culture & Society*, 35(3): 13–33.

Neyland, D. (2019) *The Everyday Life of an Algorithm*, London: Palgrave Macmillan.

Openledger (2020) 'Blockchain IoT solutions', Openledger, https://openled ger.info/solutions/blockchain-iot/ (accessed 20 October 2020).

Osiz Technologies (2020) 'Secure your smarthome with blockchain', Osiz Technologies, www.osiztechnologies.com/blockchain-solutions-for-smart-home (accessed 20 October 2020).

Parisi, L. (2019) 'Critical computation: digital automata and general artificial thinking', *Theory, Culture & Society*, 36(2): 89–121.

Peeters, R. and Schuilenburg, M. (2021) 'The algorithmic society: an introduction', in M. Schuilenburg and R. Peeters (eds) *The Algorithmic Society: Technology, Power, and Knowledge*, London: Routledge, pp 1–16.

Pegus Digital (2022) 'The progression of AI: from supervised to unsupervised learning', Pegus Digital, https://pegus.digital/the-progression-of-ai-from-supervised-to-unsupervised-learning/ (accessed 3 March 2022).

Rees, A. and Sleigh, C. (2020) *Human*, London: Reaktion Books.

Reigeluth, T. and Castelle, M. (2021) 'What kind of learning is machine learning?', in J. Roberge and M. Castelle (eds) *The Cultural Life of Machine Learning: An Incursion into Critical AI Studies*, London: Palgrave Macmillan, pp 79–115.

Rumelhart, D.E. (1998) 'David E. Rummelhart', in J.A. Anderson and E. Rosenfeld (eds) *Talking Nets: An Oral History of Neural Networks*, Cambridge, MA: MIT Press, pp 267–91.

Sejnowski, T.J. (1998) 'Terence J. Sejnowski', in J.A. Anderson and E. Rosenfeld (eds) *Talking Nets: An Oral History of Neural Networks*, Cambridge, MA: MIT Press, pp 315–32.

Shaw, A. (2018) 'Will blockchain transform the art market?', *Apollo*, 22 September 2018, www.apollo-magazine.com/will-blockchain-transform-the-art-market/ (accessed 8 July 2019).

Shen, Z., Yang, H. and Zhang, S. (2021) 'Neural network approximation: three hidden layers are enough', *Neural Networks*, 141: 160–73.

Simmel, G. (2010) *The View of Life: Four Metaphysical Essays with Journal Aphorisms*, trans. J.A.Y. Andrews and D. Levine, Chicago, IL: University of Chicago Press.

Simmel, G. (2017) *The Sociology of Secrecy and of Secret Societies*, Milan: Simplicissimus Book Farm.

Srnicek, N. (2016) *Platform Capitalism*, Cambridge: Polity Press.

Steeves, V. (2020) 'A dialogic analysis of Hello Barbie's conversations with children', *Big Data & Society*, 7(1): 1–12.

Striphas, T. (2015) 'Algorithmic culture', *European Journal of Cultural Studies*, 18(4–5): 395–412.

Suchman, L. (2019) 'Demystifying the intelligent machine', in T. Heffernan (ed) *Cyborg Futures: Cross-Disciplinary Perspectives on Artificial Intelligence and Robotics*, London: Palgrave Macmillan, pp 35–61.

Swabey, P. (2022) 'Why edge computing is a double-edged sword for privacy', *Tech Monitor*, 23 February 2022, https://techmonitor.ai/policy/privacy-and-data-protection/privacy-on-the-edge-why-edge-computing-is-a-double-edged-sword-for-privacy (accessed 3 March 2022).

Swartz, L. (2018) 'What was Bitcoin, what will it be? The techno-economic imaginaries of a new money technology', *Cultural Studies*, 32(4): 623–50.

Taylor, P. (2021) 'Insanely complicated, hopelessly inadequate', *London Review of Books*, 43(2): 37–9.

Thrift, N. (2006) *Knowing Capitalism*, London: Sage.

Turner, J. (2019) *Robot Rules: Regulating Artificial Intelligence*, London: Palgrave Macmillan.

Waters, R. (2018) 'FT big read: Artificial intelligence', *Financial Times*, 10 October 2018, p 9.

Werbach, K. (2018) *The Blockchain and the New Architecture of Trust*, Cambridge, MA: MIT Press.

Whyte, A. (2018) 'Can anyone bury BlackRock?', *Institutional Investor*, 1 October 2018, www.institutionalinvestor.com/article/b1b672fxttfp1l/Can-Anyone-Bury-BlackRock (accessed 16 November 2020).

Whyte, A. (2020) 'The relentless ambition of BlackRock's Aladdin', *Institutional Investor*, 20 May 2020, www.institutionalinvestor.com/article/b1lprrf5v41rz2/The-Relentless-Ambition-of-BlackRock-s-Aladdin (accessed 16 November 2020).

Woodall, A. and Ringel, S. (2020) 'Blockchain archival discourse: trust and the imaginaries of digital preservation', *New Media & Society*, 22(12): 2200–17.

Wu, J. (2020) 'Edge AI is the next wave of AI', *Towards Data Science*, 19 April 2020, https://towardsdatascience.com/edge-ai-is-the-next-wave-of-ai-a3e98b77c2d7 (accessed 3 March 2022).

Zuboff, S. (2019) *The Age of Surveillance Capitalism: The Fight for a Human Future at the New Frontier of Power*, New York: Profile Books.

Zou, M. (2020) 'Code, and other laws of blockchain', *Oxford Journal of Legal Studies*, 40(3): 645–65.

Index

Page numbers in *italic* type refer to figures.

Printed and bound by CPI Group (UK) Ltd, Croydon, CR0 4YY

16/04/2025

14658339-0005